IRA B. CORTEZ

1980

ACCIDENTAL
OR
INCENDIARY

Second Printing

ACCIDENTAL
OR
INCENDIARY

By

RICHARD D. FITCH

Sergeant, Baltimore County Police Bureau
Baltimore, Maryland

and

EDWARD A. PORTER

Corporal, Baltimore County Police Bureau
Baltimore, Maryland

CHARLES C THOMAS · PUBLISHER
Springfield · Illinois · U.S.A.

Published and Distributed Throughout the World by

CHARLES C THOMAS ● PUBLISHER

Bannerstone House

301-327 East Lawrence Avenue, Springfield, Illinois, U.S.A.

© *1968, by* CHARLES C THOMAS ● PUBLISHER

ISBN 0-398-00582-6

Library of Congress Catalog Card Number: 67-27922

First Printing, 1968
Second Printing, 1974

Printed in the United States of America
T-1

ACKNOWLEDGMENT

Medical discussions and terminology
courtesy of John G. Orth, M.D.

PREFACE

WE KNOW it would be of interest to the reader to know something about the backgrounds of the authors and how they came to write *Accidental or Incendiary*. As any law enforcement officer knows, men who want to break the law always manage to find each other and get together for the common purpose. Other men with a common interest are no different and will often get together to form a successful business by combining different perspectives on one common interest.

This was the case with the authors of *Accidental or Incendiary*. Sergeant Fitch and Corporal Porter, being connected with law enforcement, were assigned to the same patrol car for a number of years. Day after day, night after night, in winter, summer, and in all kinds of emergencies, you get to know your partner better than his own family. Each man gets to know the other's fears, likes, dislikes, family situation, and problems. It did not take long for us to realize we both "shined" when on a case involving a fire, even to the extent of working overtime without pay on arson cases. There was never any question about whose car to use or how long we were going to work, as we were both obsessed with solving the case.

Combining our different backgrounds gave wide coverage to the types of fires we could investigate. When on the scene of any type of fire it was always found that our combined backgrounds were the greatest asset.

Edward A. Porter, with his years of experience as a volunteer fireman, a position at which he reached lieutenancy, was most valuable when first arriving at the scene to investigate a fire. He would often tell at a glance the general area where a fire started; also, he was personally acquainted with most of the firemen and could supplement this knowledge with their information. Tracing the fire to the exact location it started however, took the efforts of both investigators. If an electrical appliance or other electrical cause was suspected, Richard D. Fitch, an electronics graduate, would engage his knowledge. His ten years experience in the automotive industry was invaluable when a motor vehicle was involved in a fire. Information on small boat fires and ship fires also comes from author Fitch who spent four years in the United States Navy in the early 1940's and many times went through the Navy fire-fighting school.

On the more serious side of investigating fires, it was often found that a human was the victim in a fire and the administration of first aid was necessary. This is an area where author Porter is well versed, having spent six years in the National Guard where his duties were in the medical field. He also had extensive training in the Baltimore Fire Department, as well as the training we both received in the Police Department, and is considered an expert in first aid.

The authors have often discussed the many books and literature they have read on fire investigation. It was during one of these discussions that the need for a book on investigating fires of all types, written by men who were there and who had firsthand knowledge was decided upon. The book had to be written in plain terms so the prospective investigator would feel he was there with us, smelling the acrid smoke and splashing through the puddles left by the fire hoses. No big words or fancy phrases to confuse the reader, just man-to-man talk and raw, on-the-scene pictures. We decided this was what was needed in the fire investigation field and this was when we conceived *Accidental or Incendiary*. The next few years were spent in extensive research and the recording of every fact obtainable pertaining to the subject. We feel that the end result was well worth the effort and hope that each reader will derive a measure of knowledge that will be an asset to him in his chosen field.

CONTENTS

ACCIDENTAL
OR
INCENDIARY

Chapter One

INVESTIGATING FIRES

INVESTIGATING a fire can be a complex operation, whether it be a small brush fire or a large building fire. In either case, upon completion of the investigation and filing of the report, the investigator is expected to have arrived at a reasonable explanation as to how the fire started. Arson squads, insurance companies, fire prevention bureaus and statisticians will read the investigators report to learn how the fire started. It is for this investigator who, in many cases, is not a specialist in fires that this book is written.

While the exact cause of many fires is never determined, if the investigator makes a complete investigation and determines the area where the fire started and mentions a possible cause, he will be "off the hook." If an investigator is not sure of the cause of a fire, he can make his initial report without committing himself and continue his investigation later. It is because of this difficulty in pinpointing the cause of many fires that we stress the importance of arriving at the scene of the fire as soon as possible.

INITIAL ACTIONS OF INVESTIGATOR

The police and the fire investigator usually arrive on the scene while the fire is still in progress and all men and equipment are still on the scene. This is the best time to investigate a fire because the firemen, on-lookers and other witnesses can be questioned on the scene. Of course, the first thing the investigator wants to learn is where the fire started. He can then determine what type of fire it is and whether arson is involved. It is important to determine these facts immediately because they will guide the course of the investigation. The investigator can now begin by questioning any occupants who were in the building at the time the fire started. They can usually tell where the fire seemed to be coming from and, in many cases, tell how it started. This information can then be clarified by the investigator. However, it is not always that easy to get to the origin of a fire. The investigator will often have to enter the burned building and search the ruins to determine where the fire began. He can ask the first firemen to arrive where the fire was burning the fiercest when they arrived and how far it had progressed. This information and the locating of a heavily charred area will give some indication as to where the fire began. This heavily charred area, of course, is an area that is burning, or has already burned, upon arrival of firemen and not an area that was consumed after the fire was in progress. After locating this area, the investigator then can look first for the easiest solution—*accidental* fire.

COMMON CAUSES OF ACCIDENTAL FIRE

An indication of accidental fire could be a charred space heater —oil, coal or wood. Check the wiring in the area and look for a badly burned fuse or electrical outlet box. In this connection it may help to determine if the electricity was operative after the fire started. If it was not, it is reasonable to assume that one of the first things to burn was an electrical circuit. In cases where there was no electricity (barns, sheds, garages and some houses), an overturned lantern or heavy charring in the area where the means of lighting was located may be a good clue as to how the fire began.

Do not discount electric motors as a cause of fire even though they smoke, give off a distinct odor, and may blow a fuse before they do any damage. A motor in a congested area, especially if it is oil-soaked and dirty, can readily start a fire. In this connection again it would be good to talk to anyone in proximity to the fire in the early stages to determine if the unmistakable odor of an electric motor burning was present.

In severe weather one of the common causes of fire is over-heated heating plants. The investigator can examine the walls around the chimney and flue outlet, also look for any trash or

FIGURE 1. In this picture you see the floor joists of the first floor of a frame house as you would see them while standing in the basement. As you can see looking up in this area the flooring is burned away completely. The alligator pattern burned into the wood joists is deeper in the area near the wall. You can see in the picture some of the debris that has fallen through from the first floor. The firemen had already started their clean up work and had just about removed everything that was left in the room above where this picture was taken. By questioning the firemen it was found that a sofa had been located above the hole in the floor. A check in the debris outside of the home revealed one badly burned foam-rubber type sofa. It was later concluded that a cigarette was dropped on the sofa causing the fire.

FIGURE 3. This picture shows a large abandoned barn on fire. The open area in the right side of the picture is due to people having stolen boards off the barn. You can see many loose boards lying on the ground below this area. There was no great amount of hay in the barn at this time. The whole farm was vacant. There was no heat or electricity in the barn at the time of this fire. This fire was found to be of incendiary nature.

building such as the chimney or the peak of the roof, but it can strike anywhere and move through the building starting a fire anywhere inside. If it strikes the electrical wires entering the house, fire can result at any outlet box or along the wires in the house. In the case of field fires, a high tree will usually be struck and damaged, having the bark or a limb torn off. Also, the tree may be burned at the top and the trunk not burned at all, since fire does not burn down a tree as well as it burns up a tree. The same can be said of field and forest fires. The fire will spread up a hill faster than it will burn down a hill, helping the investigator determine where the fire began. Always remember to question witnesses even in the case of a suspected lightning fire in an isolated area. A person a considerable distance from where the lightning struck may be a good witness.

BE THOROUGH

After a few on-the-scene investigations you will find it is not too difficult to locate the room in which a fire started, come up with some explanation as to how the fire started, and be able to make a report that will be satisfactory to all concerned. Being too hasty in your investigation, however, can lead you to believe a fire started in a room or section of a building in which it did not but which just burned more completely.

Such a case was where sheetrock wallboard was used in a room where a fire started, but an adjoining room, where the burning was more complete, had one-eighth-inch plywood paneling on the walls. In this case, in the room where the fire started the walls were blackened but still intact and the furniture was burned. The adjoining room was completely gutted and the studs were exposed. The studs are the 2 x 4″ uprights to which the wallboard is fastened and the plywood burned so completely there was no sign of any wallboard having been on the studs. It gave the appearance that the house was just never finished. Closer examination of the studs showed the ends of small finishing-nails protruding. These small, headless, finishing-nails, of course, could have been used only to hold a thin flimsy wallboard and this was soon confirmed by talking to the tenants who lived in the

FIGURE 4. Here you are standing inside a building that has been gutted by fire. The firefighters are pulling pieces of burned wood off of the wall studs to be sure that the fire is completely out. Whenever possible the investigator should examine the wall before this type of work is done. It will be much easier for him to determine which way the fire burned if all of the pieces are in their proper places. As you can see by the lower part of the studs in front of the firefighters where they have pulled the facing boards off the studs are not burned. The studs were protected from the fire by the facing boards. Without knowledge of the facing boards the investigator could be confused by these studs.

house before the fire. The foregoing example shows the importance of taking into consideration the fire-resisting qualities of all the materials used in a building that has had a fire: the sections that did not burn as well as those that burned completely.

Many remodeled buildings will be found to have what is known as a false ceiling. This is a ceiling built below the original one either to cover it up or to make a modern appearance by creating a lower ceiling. When this is done, it is not uncommon, especially for do-it-yourselfers, to leave the old wiring in the original ceiling or just extend it to the new ceiling. A fire starting between these two ceilings can smolder for a while due to the lack of air, but will really begin to burn as soon as it finds a vent. This vent may be a hole around pipes, wires, or a support leading up through the ceiling. An investigator making a rapid conclusion will say the fire apparently started here. Of course, after firemen tear down the false ceiling, which they certainly will, it will be seen that the investigator could have made a mistake. One common type of business that is continually remodeling to impress the customers is a tavern or restaurant. When a fire in one of these establishments starts, especially after closing, it is not uncommon for it to burn to the ground. There are the decorations, tables and chairs, and a greasy kitchen all of which add fuel to the fire. The common causes of this kind of fire are kitchen equipment being left on or a customer's cigarette left burning. After making an investigation at the scene, the best way to reach a conclusion in this type of fire is to question the employees and owner. They can tell you what was left on in the kitchen and what types of repairs were made recently or were needed. If you can locate any of the customers who were in the building before the fire, they may have noticed a burning odor or a faulty electricical fixture or piece of equipment. In this type of fire, once you have ruled out arson, whatever conclusion the professional investigator arrives at, using the clues from the ruins and his other investigating, will be sufficient to make a satisfactory report. There are many fires the causes of which have never been determined. Even the most experienced investigator can be stymied, but will still make an intelligent report.

FIGURE 5. The value of a good photograph at the scene of a fire is clearly evident in this picture where the fire in a resturant kitchen first appeared to the investigator to have started on the shelf because nothing on the table below the shelf was burned. The authors using flashlights in the darkened room were unable to get the overall view that the camera caught using a flash attachment. The photo clearly shows the blackened area on the wall leading down to a deep fryer that was left on; since it was made to produce and withstand heat it did not show any unusual signs of having been on fire. In most buildings where there has been a fire, the electricity will have been turned off for safety and the protection of the firemen and an investigator will never get as much light on a given area as is produced by the flash bulb on a camera. This also brings out the importance of having the pictures developed as soon as possible so you can return to the scene, before it is disturbed, to further examine anything you may notice in a photograph that you failed to see when you were at the scene.

Another common type of fire in a home or restaurant is the kitchen fire which seems to have started on an overhanging shelf or cupboard. You may feel sure the cook left a cigarette on the shelf when he went home and that it started the fire, only to find after questioning him that he does not smoke. Even if he did smoke, a cigarette left burning on a shelf would more than likely go out before starting a fire, unless the shelf was covered with paper or some other easy-burning material. A better approach would be to examine the equipment under the shelf or cupboard and determine what may have been left on by examining the position of the switches and jets. This equipment will not necessarily have any appearance of having been near the fire since it is made to produce heat and after being left on all night would become hot enough to start a fire many inches above it. We can also examine the possibility of towels or clothes being strung across the kitchen to dry and left to dangle over cooking equipment. This could start a fire in one corner of the kitchen which may do the most damage in another corner. A fire in a bedroom, where the mattress was burned, is almost always attributed to a person falling asleep with a cigarette in his hand, but the authors know for a fact that a bed lamp or a lamp in which the bulb can come in contact with the bedding is a fairly common cause of bed fires. This need not be a lamp with faulty wiring but merely one in which the heat from the bulb was in close proximity with the bedding. While we do not discount the cigarette as a cause of fire, it is not to be overrated.

Explosions

Often a fire will be accompanied by one or more large or small explosions or an explosion will be accompanied by fire. It is important that we determine whether the fire was a result of the explosion or the explosion was caused by the fire. If an explosion preceded a fire, besides the broken windows, overturned furniture and an indefinite path of the fire, the flash may have started fire in other sections of the room or building, depending on the size and type of explosion. Curtains and other flammables may be burning in many parts of the building and no definite path of complete burning can be determined from any one location. On the other hand, some explosions can blow out a fire that was already burning and caused the explosion. In this case, you will have a charred path of the fire leading to the area where the explosion was. Sometime after a fire is in progress there will be a sudden combustion caused by a window breaking or some other ventilation or draft being created and this will give the appearance of an explosion. This is the reason firemen do not stand in front of a door or window before opening it to enter and extinguish the fire.

When an investigator arrives at the scene of a fire and learns there has been an explosion, he should consider what type of building is involved and he will be able to narrow down the possibilities as to what caused the explosion. Was it a residence? a store? a warehouse? or a factory? If it is a residence and there was considerable damage from the explosion but not necessarily much fire, we can suspect whatever type gas was used for cooking or heating. If there was a leak in a gas pipe or fixture you may be able to find it easily as it may still be burning like a blow-torch. As a matter of fact that may be the only fire in the building. Determining what ignited the gas leak might not be as easy as finding it. The open flame or spark that ignited the gas can be in an entirely different section of the building and the blast centered in still another part.

Explosions Involving Heating Plants

While furnace explosions are common with gas, oil or coal, there is seldom much damage other than dislodging the heat and smoke pipes and covering the furnace area with soot. This is due to the safety features built into modern heating plants. In the case of a coal furnace explosion, the cause can usually be traced to a faulty flue or draft or the owners neglecting to burn off sufficient coal gas before banking the fire. In an oil-burner fire there can be a small to moderate explosion when the burner fails to ignite the spray of fuel and the spray hits the hot furnace. This is a limited explosion, because if the fuel fails to ignite the burner should automatically shut off. Similar safety devices are incorporated in gas furnaces where the cycle will stop if the pilot light is out. Most space heaters do not include these safety devices and are outlawed in many areas, primarily the wick and pot-type kerosene and oil burners. If a pot-type oil-burner is suspected of causing a fire or explosion, you might check the mechanism that controls the flow of fuel to the burner. Some of these burners have elaborate mechanisms to control flow of fuel and one faulty valve can cause the fuel, which is usually gravity-fed, to flow uncontrollably into the burner. While this type of oil burner is usually considered a space heater, it is often permanently set in a room, with the fuel tank located in another room, and metal tubing leading from the tank to the burner. In this case, a leaky tank or fitting on the tank or at the burner can be suspected of having caused a fire. Even if the tank is located outside the house, it is possible for a leak at the tank to run along the outside of the tubing to the heater and cause a fire or explosion.

Other Types of Explosions

Since the generally accepted definition of an explosion is a sudden release of pressure, you can see that there need not be any fire connected with it at all. On the other hand, heat can cause an explosion in which there is no combustion at all. In the case of a hot water heater or steam cleaning equipment, a faulty safety valve could be responsible for a considerable explosion. Compressed air tanks and equipment used in cleaning and laundry establishments may have explosions without resultant fires. In investigating explosions, consider the amount of damage done and determine if that amount of damage could have been done by what you suspect as the cause.

EXAMPLE: Suppose a two-car garage at the rear of a residence was completely leveled by an explosion and one or two cars inside were completely destroyed. Upon questioning the owner he states that the gasoline stored in the garage for the lawn mower probably exploded. Your experience will tell you that this is not a normal situation especially if you arrived on the scene minutes after the alarm was received and found total devastation in the area of the garage. Even if gasoline is more powerful than dynamite, the average citizen, by his own conscience, does not store more than perhaps five gallons. Even this is stored in a metal container and is not in the atomized state in which gasoline is most potent. In a case like this you can start searching the ruins for a substance or residue foreign to a residential garage.

CONCLUSIVE REASONING FOR EXPLOSIVES: This same reasoning can be applied to all investigations of explosions. Any suspicious odors, residue, color or smoke alien to the type of explosion you suspect should give your investigation a slant toward possible *incendiary* cause of the fire. Even if an explosion is the type common to a certain industry or material you will have to make sure it was not caused deliberately or through neglect. There are set regulation in all areas governing the use of dangerous materials and equipment and these could have been violated deliberately or through neglect. This will include storing and handling flammables and explosives. When investigating explosions, consider the course the explosion has taken. The blast will follow the path of the least resistance. So in a residential explosion in a house made of frame, for instance, with a peaked roof and a finished attic, it is possible to blow out all the walls without damaging the heavy reenforced roof. On the other hand, a brick or stone building with a flat roof can have the roof completely blown off by an explosion. A reenforced concrete building with an internal blast will have all the windows shattered and extensive damage inside. Some sturdy buildings will have a hole blown in the wall if the pressure from the blast can find no other area of weaker resistance. Block foundations usually give way easier than poured concrete and if a blast is severe enough in a basement a section of the blocks will fall and release the pressure. Some buildings that are used to store explosives are purposely constructed to control the direction of a blast in the case of an explosion.

THE HUMAN TORCH

Added to the many types of fires already mentioned is probably the one that is the most serious from a humanitarian viewpoint because it involves human life. This is the case where a human torch is involved. A human torch, of course, is a person who becomes engulfed in flames, often while fully awake and conscious. This may be due to carelessness or neglect or may be deliberately done by some other person as a prank that backfired. It is not one of the popular forms of suicide in the United States; however, in some foreign countries it is gaining popularity as a method of focusing attention on a particular cause. Some persons are more apt to set themselves afire than others. A man who works with flammable liquids or gets his clothing oil-soaked may go for years without an accident, and then become careless. A mechanic cleaning parts in a can of flammable liquid, or a painter with paint-covered coveralls mixing paint with a lit cigarette in his mouth is a potential human torch.

When you investigate a fire where a person was burned and that is the focal point of the fire you may not be able to question him and will have to make your investigation without this advantage. If the victim was alone at the time of the fire, you can begin by determining what type activity he was engaged in. As soon as you get this information you can isolate the probable cause. If the fire occurred in an auto-repair shop, a paint shop, home, or a factory, you can determine the possible cause. If the fire occurred in a home, the most likely place is in the cellar. The cellar is where men mix their paint, store their various gasoline motors and do any dirty work that requires cleaning mechanical parts. This is very dangerous because the cellar, being below the ground, does not afford proper drain off for the heavier-than-air vapors even if the usually small cellar windows are open. A quick look around by the investigator will give the clue as to what happened. You may see a partially dismantled power mower or a wash basin on the floor with some blackened machine or motor parts in it that were being cleaned. If the victim was doing some remodeling work such as laying floor tiles, or painting or varnishing, this probably contributed to his setting himself afire. All you will have left to determine is what set it off. If the victim had the object in his hand he probably threw or dropped it. See if there is a cigarette lighter or a blow torch laying on the floor. If not, check any pilot lights or other spark-producing equipment for traces of a flash having originated there. The same unfortunate chain of events that causes human torches in cellars of their homes can be applied to private garages, especially in the winter.

While the family garage is not normally in the cellar of a private home, a subject wanting to do some work during the winter will duplicate the cellar conditions by working in the garage, closing the doors and windows tight to keep warm, and even putting a small heater in the sealed garage. All this while the gasoline-laden family car is also in the garage. In a fire of this type the car may be burned considerably, which would lead you to believe it was involved in starting the fire, but this is not necessarily so. If the hood of the car is up, you can assume the victim was working on the car. Try to determine just what he was doing to the car. Are there parts missing from under the hood? If the carburetor or the fuel pump is missing, look around for it. When you find the missing part, if it is on the floor as if dropped or thrown, or if it is soaking in a receptacle on a work bench, you may more easily determine to what extent the part was involved in starting the fire. When fuel lines are disconnected from a gasoline engine, a certain amount of fuel will usually leak out onto the floor and the hands of the victim. About this time, if he decides to take the cigarette from his mouth because the smoke is getting in his eyes, or picks up a cigarette to reward himself for successfully removing the part, you will have your answer as to how the fire started. In most cases involving a human torch the victim will have gotten some type of accelerator on himself or his clothing. This can include vapors in his clothing. Of course, a person can be caught in an explosion or flash fire and be severely burned but the victim in this case would not necessarily be the cause. The investigator can consider the material in the clothing. In flash fires or explosions the clothing of a victim will sometimes afford protection and he will only be burned on the exposed parts of his body.

The reaction of a person who suddenly finds himself engulfed in flame will differ depending on his personality and any training he may have had. The natural impulse is to run, which is the worse thing he can do as this only fans the fire. A person who can compose himself would most likely roll on the ground or try to tear off his clothing. The victim's trying to flee the fire can make your investigation complicated because they will start other fires in their travels and you will have to try to backtrack their movements to tell where the fire began.

FIGURE 6. We all know from our own experience that a wool sweater will produce a spark when it is rubbed or agitated a little. The citizen who owned the sweater in this photograph decided to wash it in a pan of gasoline at the bottom of his outside cellar way. The pan of gasoline with the wool sweater in it was completely outside the cellar when, as the owner was beginning to dip the sweater up and down in the pan, it burst into flames. The man said he was not smoking at the time and could not explain what caused the fire. While it is unknown to the authors if gasoline saturated wool will cause a spark the same as when it is dry it was reasonable to assume that since the wool sweater had just been placed in the gasoline it hadn't become completely saturated and the first movement of the sweater in the pan by the owner released the spark at a time when the proper amount of air was mixing with the vapors to cause the detonation. The owner of the sweater did not panic and managed to extinguish the fire with a minimum of damage and minor first degree burns on his hands.

Juvenile Torches

In the case were young children are involved, you can suspect that they were being inquisitive or disobedient. Children are not likely to have been involved in any repairs or cleaning as in the case of adults. Some very young children have been known to take to the smell of gasoline and other flammables, even to the extent of spilling some from a container to watch it while inhaling it. Other children (and some adults) will light a match to see what is in an empty or partially filled tank or container. When the investigator arrives on the scene and finds a container or tank with the cap missing and possibly some evidence of burned matches, he can determine what was in the tank and close the case. Some parents thoughtlessly dress their children in highly flammable costumes to celebrate a birthday or holiday. An investigator who arrives on a scene where a child has been burned and observes a celebration of some kind was in progress will most likely find the celebration had something to do with the catastrophe. Costumes, decorations and candles can be considered prime suspects.

Dump Fires

Dump fires are generally attributed to natural combustion. But very often this is not the true cause. If we were to stake out many dumps we would find that trucks are dumping loads of burning trash on the dumps. In some cases, we may find that the dump owner is setting the fire himself, as the more trash that burns the more space he will have to dump other trash. It is also possible that children playing around a dump will set fire to it. And, in some cases, the rats living in the dump will help spread the fire.

Often, the fire will begin burning deep down inside the pile of trash. It may smolder and spread underneath where it cannot be seen for a long period of time and then break through the surface. In many cases, rat holes help the fire get air and assist it in spreading. It is possible that you will find a fire on each end of the dump and they will appear to be two different fires but actually they will be all part of the same large fire which has been spread underneath the dump.

FIGURE 7. This picture shows us just a small part of a very large dump that is burning. In a fire of this nature you will find that you have many different types of items burning. In just this small section you can see paper, cardboard, wood and all different types of metal drums and containers. It must be remembered that all of these containers may not be completely empty and may contain some flammable materials. This could be of great danger to the firefighters and the investigator. When piles of trash like this burn, much of the actual burning will take place underground, out of sight. The investigator could feel that he was away from the fire and then tread on an area that is burned out underneath which would break through under his weight causing him to fall into an area of hot ash.

Due to the many hot spots and holes, it is a *must* that the investigator wear boots.

When investigating the dump fire, the owner, his employees and anyone living within sight of the dump should be questioned. Also, it is important to question anyone who has a view of the access road to the dump, as they can give information as to who was seen coming and going from the area. They may also give you information as to trucks carrying hot loads into the dump. A check of fire department records will tell you if this dump has many fires. We can then check to see if the dump is in violation of any local laws. If there is a record of past fires, we may then decide to stake out in the area.

RECONSTRUCTING THE FIRE SCENE

It has been the experience of the authors to arrive on a fire scene after the firemen finished their work. Their quick action promptly extinguished the fire and confined the damage to the one room where the fire started. After they threw all of the furnishings and cabinets out in the yard they tore off the wall covering and ceiling, leaving only the unburned wall studs and ceiling joists for us to examine. Under these conditions, determining what room the fire started in was the best we could do, and it didn't take a trained fire investigator to make that decision.

In order to isolate the cause of the fire to a more definite area and object in the room it was necessary to reconstruct the scene. This can either be done outside where the firemen threw the contents of the room, or by bringing the contents back into the room, when practical, and putting them in their original places with the help of someone familiar with the scene before the fire. In one case we completely reassembled the cabinets and fixtures from a unit kitchen in the back yard. This clearly revealed the course of the fire and pinpointed the cause.

If reconstructing the scene for some reason is impractical, similar results can be obtained by examining the burned objects, then drawing a sketch of the burned building noting the location of each object on the sketch. The extent and sides or part burned should also be noted. In any event, photographs should be taken

FIGURE 8. This picture is a pile of rubbish in the back yard of a residence that had a fire of undetermined origin and the investigator was delayed in arriving at the scene. Even if something of a suspicious nature was located in sifting through the debris we would have to find out as close to exactly where it came from in the house as possible. You will then find out that the firemen during their clean up operation do not record every shovelfull they throw out. The moral of this picture is then to "arrive on the scene as soon as possible."

of a complete fire scene if you intend to continue your investigation. These photographs can be helpful to determine the course of the fire when compared with your sketch.

After you have reconstructed the fire scene you may determine that the fire started in a particular corner of a room where now there is nothing but the bare floor and walls. The person who will have the answer to this is the person whose aid you enlisted in the first place, because he was familiar with the room before the fire. He will be able to recall that there was a box of trash or excelsior in the corner that was completely burned and the ashes just swept out by the firemen leaving no trace.

With this information you can pretty well complete your report. In an *accidental* fire it will be sufficient to say that the fire started in a box of trash or excelsior in the corner of the room, unless you have some reason to make a case out of it.

Reconstructing a scene and finding nothing in the area where you believe the fire began is particularly apt to happen with industrial fires. Packing crates, cardboard boxes, pallets of other flammable materials are often stored in large quantities. By the time you arrive, these materials could be completely consumed and unless you ask someone what was stored in a particular area, you may never know.

Tracing the route of a fire by examining the burned objects that were in its path is an excellent aid to locating the origin of a fire. If you find, for instance, a window frame blackened and it appears the curtains and blinds helped spread the fire, you will want to know what was in front of or against the window. When the resident of the house helps you bring the sofa back in and places it against the window and your examination of it indicates the cushion and back burned, but not the bottom or the part that was against the wall, the simple conclusion is that someone ignited the sofa cushion.

In areas where fire protection is good you will find most of the investigations easy to close, as in the foregoing example. Only if a fire gets completely out of hand or there are suspicious circumstances will you have to conduct a major investigation.

FIGURE 9. Close examination of this on-the-scene photograph taken by the authors might indicate arson in that the fire seemed to have its origin in two places. One against the wall and another in the middle of the floor where, according to the occupants of the house, there was nothing in that particular place. Close examination of the ashes in the area of the burned spot in the center of the floor revealed small metal objects such as axles that were identified as parts of plastic toys that were piled in a corrugated carton against the wall where the other burned area is. Behind the box in the wall was an electrical outlet with a worn line cord plugged into it.

The reasonable explanation to this was that the toy carton caught fire and toppled over spilling the burning toys into the center of the floor. The blackened copper wire from the burned line cord is along the base board and is slightly visible in the photograph. You will find it best to take many pictures of the suspected area of origin from different angles because shadows and reflections produced by the flash will obscure some objects and emphasize others.

SUMMARY

1. From the foregoing you can see the importance of arriving on the scene as soon as possible to begin your investigation. Common sense will tell you that if you arrive three days after the fire has been extinguished and the souvenir hunters, looters and local juvenile delinquents have swarmed over the scene, it will be impossible for you to conduct an investigation.

2. The importance of questioning witnesses cannot be overstressed. This has to be done as soon as possible while they are still available at the scene.

3. When properly questioned the firemen at the scene are a valuable source of information especially if upon their arrival suspicious conditions were observed such as open windows or doors, or evidence of forced entry.

4. Be wary for the common causes of fire, for the prevailing season, in the type of building being investigated.

5. Also be alert to other causes, including spontaneous combustion, or lightning if there has been an electrical storm in the area.

6. Do not be too hasty in your conclusion; experience will tell you that a fire is not always the type it appears to be. A professional arsonist will invariably make his fire appear *accidental*. Conversely, an *accidental* fire can appear to be incendiary.

7. An explosion has to be carefully investigated to determine if it was the result of or the cause of the fire. This can be determined in part by observing combustables that had been afire but the explosion had blown out the flames.

8. In cases of explosions always consider the amount of damage in relation to what you believe to be the probable cause. When the amount of damage exceeds the probable cause or there is the presence of alien odors or colors of smoke, arson can be suspected.

Chapter Two

SAFETY PRECAUTIONS

INVESTIGATION at the scene during and after the fire is a dangerous occupation and can often result in sickness, injury or even death to the investigator. If reasonable safety precautions are observed, these possibilities can be greatly reduced or eliminated. Danger to the investigator is greater than to others at the scene because he is usually preoccupied with his investigation and cannot be concerned with other activities around him or the progress of the fire. Investigators who respond to the fire alarm have the added hazard of traveling to the scene under emergency conditions. These men should realize that the safest route is the fastest, as even a minor accident can detain them long enough to set the investigation back. Generally, investigators do not carry ladders, life nets, hoses or other life-saving equipment, and should give the right-of-way to vehicles that do. Fire equipment may be responding from many directions, so be alert at intersections, you may not hear the siren because of your own.

Protective Clothing

Upon arriving at the scene, prepare yourself with proper clothing. Heavy-soled overshoes are a must because of the glass, nails and other sharp objects that will be lying around waiting for you to step on them. If the temperature is below freezing, the water from the fire hoses will freeze and add to the walking hazards. A raincoat and helmet are also good investments. Even if the fire has already been extinguished, the raincoat will keep the acidy odor left after the fire out of your suit or uniform.

Dangers

Conducting an investigation in a burned-out shell, where the fire was put out a day or more previously, does not mean the

FIGURE 10. This building was being remodeled when fire struck doing a great amount of damage. The window frame on the right side of the picture is blackened at the top. This would indicate that a large amount of heat and smoke had escaped from the building at this point. This type building is a danger to the firefighters and the investigator. Often there will be loose boards with nails in them which you may walk on and run a nail in your foot or trip over and injure yourself. There may be floors that are only partly laid or temporary supports that are head level. These problems are danger enough under normal conditions but even more so at night or when the area is filled with smoke.

Figure 11. In this picture we are on the first floor of a home, looking up at the ceiling. The fire started on the second floor and burned through in this area of the ceiling. Some of the ceiling was pulled down by the firefighters in their attempt to make sure the fire was out. Looking through the doorway where the fireman is standing, you can see that part of the ceiling is hanging down in the hallway. The wall in the hallway shows soot damage under the area where the ceiling was burned. The danger of pieces of the ceiling falling is one reason that the investigator should wear a helmet in his work. If a large amount of water is used by the firefighters, it is possible that a section of ceiling may fall from the weight of the water alone.

FIGURE 12. This picture shows the burned-out floor in the second story of a home. You can look through the holes and see the room on the first floor. This fire started on the second floor in this room. The floor being so badly burned would suggest that the fire started in this area. Badly damaged floors such as this present a danger to the investigator. If you must walk on a badly damaged floor, then you should try to walk where the floor joists are. It is best never to work alone in a damaged building.

FIGURE 13. This picture presents great dangers to the fire investigator. There is a large beam running from the lower left side of this burned-out house to the upper right side. This beam is supporting other boards and is ready to fall at any time. There is a large board in the right side of the picture, running from the second floor to the chimney. This board is also ready to fall. The chimney itself is a great danger. The heat from the fire will, in most cases, damage the mortar in the chimney. In the cases of an old house, age itself will often have taken effect on the mortar. The investigator should not endanger his life by working under this condition. At this point it is no longer a matter of life and death, and the investigator should allow these dangers to be removed before he starts his work.

walls, floors and ceilings are safe just because they did not collapse or cave in during the fire. Cooling and contracting can cause walls to buckle, staircases to shift or a whole roof to cave in. Plaster ceilings are especially dangerous when they have been saturated; it may be days before they come down. Before entering a building that has been burned, the investigator should examine the outside walls and the joists under the floor on which he is going to walk. The ceiling and roof can then be examined to ascertain if they will remain in place long enough to complete your investigation. While probing through the ruins care should be taken not to remove a piece of lumber or other material that may be acting as a prop and bring the whole building down on you. Remove debris from the top of the pile first; then work your way down, handling each piece with care and watching for exposed nails, broken glass and other injurious matter.

Health Hazards

In a large fire, where chemicals or other noxious materials were burned, do not remain in the building for prolonged periods without resting and getting some fresh air. Prolonged periods of breathing poisonous vapors can have far-reaching effects which may not become apparent for many years. The same, of course, holds true in cases where the fire is still in progress. If the heat or smoke becomes so intense that your life or health may be endangered take what ever precautions may be necessary. In most fires the electricity will have been disconnected, but this is not a definite rule and with your wet shoes on the wet floor you should be able to complete any electrical circuit your hands or body comes in contact with. This, of course, can be fatal, so when in doubt check with the utility company. If electricity is still connected to a burned building and it is hampering your investigation, have it disconnected or turned off. The authors have found utility companies prompt and cooperative in any situation. If you are working at night carry adequate light so you can see where you are going. The light should be placed in a given area so you can work with both hands free.

An Aide For Safety

When working in a building that has been gutted or is still burning, it is best that two investigators work together and stay together. Both men can then be alert to signs of danger that may require them to evacuate the building in a hurry. If you are the only investigator, be sure that some other official on the scene knows you are in the building. At the slightest sign of shifting, settling or cracking, the investigator should get out of and away from the building. As a matter of fact, an investigator in a dangerous building should at all times know the fastest way out and the second fastest way out. When it becomes necessary for you to evacuate a building because of danger of collapse, move completely away from the building and warn others in the area who may be in danger.

Firemen Can Be Dangerous

Investigators working in a building where the fire department activity is still in progress have the added problem of staying out of their way. When the firemen are tearing down a wall or throwing debris out a window, it is not usually foremost in their minds that an investigator may be on the other side or down below. Also stay clear of the high pressure hoses that are sending a stream into the building through a broken window as particles of glass will be coming in with the water. Remember that the firemen are just as preoccupied with putting out the fire as you are with investigating it and you will have to look out for yourself.

Below-Grade Hazards

When you find it necessary to enter the cellar or basement of a building you will probably see a foot or more of water on on the floor. Wearing boots is not the only requisite for wading into this water. First, be sure the electricity leading into the building has been disconnected by the power company, then examine the floor joists above you, then be sure there is no gas leakage or other vapors accumulated in the cellar. Next check for your nearest secondary exist, which may be a window as most cellars are lacking in doors. A flooded cellar is one of the worst places to be trapped, so advise other officials on the scene before you enter the cellar. When walking in the flooded area you will have to walk very carefully, as you may trip on an unseen object or fall into a hole you could not see. Also there may be flammable liquids floating on top of the water from a fuel line, tank or a container that was in the cellar. Even if the fire did not touch the cellar an investigator will invariably have to enter it to examine the fuse box and the heating plant.

Rescues

If you are employed by the department of public safety which includes the police and fire departments, your timely arrival at the scene of a fire may put you in a position where you will have to enter the burning building to effect a rescue of someone trapped or overcome in the building. You may be the only official on the scene at this time and some kind of effort by you will be mandatory. It would be foolish to dash into the burning building without first determining exactly what room, on what floor, the victim is believed to be. With this knowledge and your awareness of the danger from the heat and flames, the possibility that you may lose your sense of direction due to the smoke and excitement and the potency of the poison gas that is sure to be present and can be fatal if inhaled even in small amounts, you can prepare to enter the building. To protect yourself from these three dangers you can cover yourself with a saturated blanket, if possible, and then stay low in the building. Take a lifeline into the building with you, having someone hold the other end outside. This will enable you to find your way out if you get lost and it will be of help to other rescuers if you are overcome with smoke. Then take a deep breath of fresh air; *do not* breathe while in the building, as once you inhale in a burning building you will have to make a quick exit. If you know just where you are going you should be in and out of the building in less than two minutes.

Of course, in all cases where you must enter a burning building common sense must prevail. Even if you are the first official to arrive on the scene where persons are trapped inside and all eyes are on you, the decision of whether or not you can make it is yours. If the building is completely engulfed and persons are known to be trapped inside, it is reasonable to assume they are already dead and the suicide of a would-be rescuer will not help matters. The authors have been confronted with this situation and the feeling of helplessness cannot be described. The only thing you can do in an impossible situation is to attempt to expedite the fire-fighting equipment, by radio or by phone, and protect the area so no one else is injured.

Knowledge of your limitations and complete familiarization with danger accompanying fire is the investigators best insurance against death or injury. Especially remember the danger of poison gas. Everyone knows the obvious danger of the heat, smoke and flame that accompany a fire, however, many of the fire fatalities on record never came in contact with enough heat, smoke or flame to cause death but were killed by inhaling the poison gases emitted or generated by the various burning substances.

While most burning substances give off an obnoxious odor which you will naturally avoid and try not to inhale, the gas that is a deadly poison may not have any accompanying odor at all, or it may even have a pleasant odor.

The man who chooses to deal with fire as his career will have to learn at the onset when to inhale and when not to inhale or his career will be a short one.

SUMMARY

1. Many of the dangers to the on-the-scene fire investigator will be more apparent after your first assignment. Remember that many of the dangers will have far-reaching effects that will not be noticed for years to come. It's up to you to take care of yourself.
2. Always have all of your protective clothing and equipment with you when going on an assignment. If it is not supplied by your employer, it is a good investment in your own future.
3. Enter burned or burning buildings with caution and have an aide with you if possible. Always advise other officials at the scene that you are entering the building.

FIGURE 14. Here is a picture of the remains of one section of a sectional sofa. As you can see, it was almost completely consumed by fire. The fire started in this sofa and completely gutted the first floor of the home. A close examination of the sofa will show you that it was padded with foam rubber. The burning of this foam rubber along with other items in the home caused a very heavy black smoke. This smoke along with the heat hampered the firefighters in their work and took the lives of several children in the home. The children were overcome by the smoke and fumes before the flames ever reached them. The fire spread very rapidly because of drapes on the windows behind the sofa and the house being an old one of frame construction. The children were trapped on the second floor because the stairway was open and led directly into the room where this sofa was located.

4. Do not spend too much time in burned or burning buildings as this is injurious to your eyes, nasal passages, and lungs, and the scent will be with you for days.

5. Make sure you know where the nearest exits are in case you have to make a hasty retreat.

6. If the fire is still in progress the hazards of the investigator are increased by the presence of the firemen whose only concern is extinguishing the fire. Unless the investigator believes a crime is involved he need not enter the burning building until the fire department declares the fire "under control."

7. Probably the most dangerous area of a burned building is the part below grade. All the water from the fire hoses settles there as well as much of the debris dislodged from the upper floors. Extreme care must be exercised when working in this area. *Make sure someone knows your whereabouts.*

8. When your early arrival puts you in a position where it is necessary to risk your life to save another remember the three dangers you will meet: (1) *heat and flame;* (2) *inhaling smoke and poison gas;* (3) *loss of sense of direction.* Prepare yourself against these dangers and you can effect a successful rescue.

Chapter Three

ACCIDENTAL FIRES

ACCIDENTAL fires are fires that were not deliberately set. An investigator must be assured that a fire which appears to be *accidental* was not *incendiary* and *made to* appear *accidental*. In some cases, the clever arsonist will attempt to make the fire appear to have been started because of carelessness or some other natural cause. However, most fires are accidental and, after a reasonable cause has been found, can usually be written off as such. The definite determining of any of the natural causes of a fire listed previously can close the report on accidental fires. In many cases the fire could have been accidentally started by an occupant or workman in the building who, at first, will not admit it. This could be true of using a blow torch or other equipment that could cause fire. As soon as the investigator suspects the fire was accidentally set, he should intelligently question the suspect and confront him with the evidence immediately while he is still emotionally unnerved from the fire. The average citizen, who is not a criminal and cannot match the trained investigator, will readily admit he could have started the fire.

Small Children

Small children are often responsible for accidental fires and because of fear of their parents, do not readily volunteer information. They may be playing with matches or lighting their way in a dark cellar or attic or imitating their parents by feigning smoking or lighting the range. When this is suspected the parents should be questioned first tactfully. Then the children should be questioned with the permission, but not in the presence, of the parents. Only experience will give the investigator the ability to gain a child's confidence and get the information he needs. It will be found that some children have great composure when being questioned. It may be necessary for the investigator to put them at a disadvantage by examining their hands and clothing for evidence of their being near the fire. Once the child realizes the investigator knows his business, the case can be quickly closed. Investigators may find it difficult to bring children into an investigation for personal reasons or because of criticism from onlookers or relatives. The fact is that an alarming number of fires are started by children, either deliberately or accidentally, and the investigation must be completed regardless. When a good investigation is completed those who criticized will be apologetic and grateful and the children who were involved will gain by the experience and be better citizens by having respect for the process of law and an awareness of the dangers of fire. The parents may then come forward and admit that their child had, in the past, been caught with matches and warned of the dangers. Many children who are responsible for fires have a tendency to play with matches that the parents may not be aware of.

Personalities

When questioning adults about accidental fires, you will find yourself confronted with a number of different types of personalities. A person who is a tenant in a house may be reluctant to admit he could have accidentally burned down the landlord's house. Other types of persons will contradict every question you ask, thinking you are accusing them so you can place charges against them. This person will usually be loud and excitable and will have to be reassured that you only want to make a routine report after which he will be more cooperative. If you are reasonably certain that a fire was accidental and there has been considerable loss, property or otherwise, you can sometimes get more and better information by introducing yourself to the principals briefly and advising them you would like to talk to them as soon as they are able so you can complete your report. This is a humane approach to be used when you think it fits the situation at hand. If you have a fire to investigate where there are conflicting stories from the witnesses, remember that they are not necessarily lying. If you come back and question the same persons again, you will find yourself talking to altogether different personalities after the tension and excitement has left them. The excitement of a fire in proximity to an average person is not an everyday occurence and they will often confuse the facts if required to relate them while in their excited state. If you return to question the same witnesses a day or so after the fire, you may find an added advantage of persons talking among themselves and relating information to each other that they did not tell you because you did not ask. You may find that a person who had information was not questioned at all.

Visual Observation

When talking to the occupants of a burned house who claim they could not have started the fire accidentally because they are very careful and never leave appliances on, or overload circuits, or are careless with their cigarettes, the investigator can examine the unburned portions of the house and possibly get a different answer. Cigarette burns on the window sills, tables or piano indicate a careless smoker. If the fire started on the floor or on some object beneath one of these handy ledges, you can mention this possibility in your report with a note about the other cigarette burns throughout the house. You can apply the same reasoning to other conditions found in the unburned parts of a house such as faulty electrical wiring, antiquated appliances or heating plant or just general carelessness. Persons being questioned about a fire will be very "put-out" if you suggest they were poor housekeepers, but a check around the rest of the house will give you your answer. Poor housekeeping may have not been the cause of a fire but you can be sure it helped add to the fuel, especially if the fire was in the cellar.

Neglected Children

Another type of serious fire receiving much publicity and seeming to be on the increase, especially in the larger cities among the poorer families is the fire occurring in a flat or tenement house with small children left alone, or in the charge of a very minor child, who burn to death. These cases are justly drawing the attention of civic and welfare groups who are advocating care centers for these children. Usually, these families are victims of the times and the high cost of living. The father, who lacks education or a good trade, will have his income supplemented by his wife who takes a part-time job in the neighborhood. Of course, with conditions such as these, there would be no percentage in hiring someone to watch the children. The next best thing this lower income family can think of to do with their children is to lock them in the apartment, or maybe in one room, where they will be safe. In most instances, these children will be of pre-school age; they are old enough to be inquisitive but too young to realize danger. This is the combination that is often fatal to the children and heartbreaking to the parents, when one or both of them return from their jobs and find the fire department still at the scene.

An investigator on an assignment of this type may readily find the cause of the fire in the gutted apartment, but where to place the blame will be another problem. Was it the child's disobedience, the parents negligence or the blind society that is responsible for this increasing problem?

Delinquent Parents

Of course in the foregoing-type cases, neglect charges would not be brought against the bereaved parents unless it was to focus attention on their plight and all others like them. The preferring of charges may bring enough publicity to the condition that a remedy or partial solution may be devised that will forstall any similar catastrophes in the future.

There is, however, the case of neglegent parents where you would definitely want to prosecute. If you had a situation similar to that previously mentioned where the children were left alone and a fire occurred but, instead of the parents being at work earning the means to hold the family together, they were at the neighborhood bar or a party, with total disregard for their children, we would say this is a clear case of parental neglect and should be prosecuted.

The Old Or Infirm

Fires in which the old or infirm are the victims are common. This can be attributed to many factors. Unlike the healthy, young adult the elderly or infirm person often suffers from a dulling of the senses which would forewarn them to danger. Blood circulation of the elderly is usually poor, which will induce them to take extra measures to keep warm. When an investigator is called to a scene where an elderly person, at home alone, succumbed, he must take all these facts into consideration as they will help him to reach a conclusion as to how the fire started and why the victim failed to escape. He should inquire of neighbors and relatives as to the physical condition of the victim, and whether or not anyone periodically looked in on them. An autopsy may have to be performed to determine if the death may have been due to natural causes and, in some way, caused the fire. On the other hand, it could have been a heart attack brought on by discovering the fire. In this case an autopsy may reveal some evidence of the fire in the victims lungs.

Many old people become feeble-minded and do things which do not appear normal. One may light three or more cigarettes at a time and forget where they were left, or perhaps light the oven and forget to put the roast in it. Forgetfulness is a common trait of the old and feeble-minded and the investigator can facilitate his investigation by familiarizing himself with the mental and physical status of the victim.

Imbibers

Throughout the country there are a large number of persons who habitually partake of intoxicants and who live alone in rooming houses. These persons will indulge beyond their capacity, then find their way to their room, sometimes with help, and collapse on the bed. Of interest to the fire investigator is the fact that the mortality rate from fires among these persons is considerable. When a fire investigator is called to a rooming-house fire and his questioning of neighbors indicates the victim was of the afore-mentioned character, he will usually find the victim here

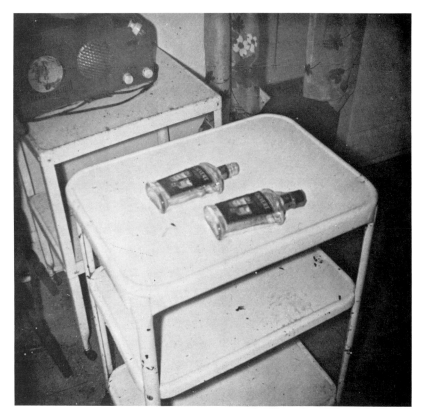

FIGURE 15. The night man at a motel saw smoke coming from one of the cabins. He called the fire department and when they arrived they found that the bed in the cabin was on fire. No damage was done except to the bed. No one was in the room at this time. The night man was questioned and he stated that earlier he had rented the room to a lone man. The man had left since then, but he did not see him go. He also stated that the man to whom he had rented the room had been drinking. The table shown here was located next to the bed. It appeared that the subject laid in bed drinking and smoking, and then left in a drunken condition, but not without leaving a hot ash or a burning cigarette in the bed. So here we have a very common cause of fires, the combination of alcohol and cigarettes.

burned to death in his bed. This will be true because the victim, in his drunken condition, had no desire to employ any auxiliary heating equipment or even to turn on a light or any other appliance. The only thing he did was to fall on the bed with a cigarette between his fingers. If the victim's fingers had been examined before the fire, burns would have been noticed between his fingers where he had fallen asleep on previous occasions holding a cigarette.

Fireplaces

The open fireplace, which was replaced by central heating plants in the home and, therefore, serves no useful purpose, is making a comeback. People who have one installed in a new home with a chimney as wide as the fireplace do it mostly for appearance. They will spend a day cutting logs or will purchase them so they can have a fire. Most of the heat produced by a fireplace goes up the chimney; the little that gets into the one room where the fireplace is located does not justify the danger of an open fireplace. You will find that people who can afford this added accessory in their new home will also be able to have highly finished floors and a thick rug in the room where the fireplace is located. In investigating a fire where you suspect a fireplace as the cause, do not discount your suspicions when a survivor of the fire tells you it could not have been caused by the fireplace because there was a screen around it. A minute spark could have gone through it. This type of fire will be hard to investigate without the help of witnesses to give a clue as to where the fire was burning in its early stages. If there is total burning you may find nothing standing but the fireplace which was the cause. You can, of course, question neighbors who have fireplaces and ascertain if their fireplaces were burning well or if they noticed any unusual downdrafts. If neighbors of the fire victims also have a fireplace and they say that their fireplace was burning erratic with downdrafts, its reasonable to assume that the fireplace in the burned home had the same conditions.

Auxiliary Heating Units

Other types of auxiliary heating units are often the causes of fire. These are usually found in basements, club cellars and spare rooms. During the cold season, the central heating plant is often inadequate to keep sufficient heat down in the cellar to make it livable and an auxiliary heater, usually of the portable type, is used. Of course, the heater would be in the way if it were placed in the middle of the floor, so it may be put in the corner or near a highly finished wall. The wall near this heater, as well as the furniture in front of it, can be the origin of the fire. This is especially true if the heater is of the type that concentrates the heat in a certain direction. A flammable object many feet from this type heater can catch fire. Also, for appearance, the portable heater may be put in an artificial fireplace which will confine enough heat to cause a fire. In a spare room that is used for sleeping, a heater may be placed near the bed to warm it up before retiring. This type of fire will be ob-

FIGURE 16. This photograph shows the top view of an electric range taken from a house fire. It shows that even though it has been burned, you can still determine which units were turned on and which were turned off. Note that the three handles that are still intact are all in different positions. When the knob is completely burned away, such as is the case of the lower left knob, then we must examine the metal shaft. In most cases, one side of the shaft will be flat, this will then be compared with knobs and or other shafts and tell us if the unit is turned on or off. This will be very important in determining the cause or eliminating some possible causes in many fires.

vious to the investigator as the bed will have been thoroughly warmed and the heater will probably still be in the position it was when it set the bed on fire.

The investigator will find that even people who are meticulous and careful about everything else will use poor judgement during the cold season to make sure their home is warm, even to the extent of turning on the cooking equipment and leaving it on with other auxiliary heaters while they sleep. As we know, the cooking equipment is functioning properly when it is being used for cooking. When it is used for heating, a purpose for which it was not intended, the open flame is usually exposed and can be hot enough to start a fire as far away as the ceiling which may be four feet above the open flame. On the other hand, it may not have been left on intentionally. The victim probably intended to turn the burners off when the house became warm enough, but was overcome by poison gas first. This can also be true of other types of open-flame heaters. In most cases the heating units and portable heaters are not connected to a flue and this causes the heat and poison gases to come into the living area. Keep this in mind when investigating a fire where persons were unable to escape or did not try. It is not uncommon to find lifeless bodies in a room that was not excessively heated. The fire in an adjoining room, however, produced enough poison gas from the material burning to be fatal to the victims. There are as many fires caused by faulty chimneys as are caused by not having a chimney.

Chimneys and Flues

Investigators working on a case where it appears as though the fire started in a wall will often find, after the firemen tear the plaster off the wall, that the chimney is on the other side. If the chimney is made of brick, one may assume there was adequate insulation to prevent a fire. However, over the years, settling of the house and the action of the elements as well as the temperature changes will cause the mortar to disintegrate. There may be dangerous cracks in the chimney, especially around the roof, and this may cause a fire in the rafters. A flue liner is the best insurance against a chimney fire; although in many areas it is not required by law. An investigator will find that most of the aforementioned types of fires are associated with extreme cold weather when inadequate heating plants are overworked and people are using every method at their disposal to keep warm. Of course, the direct cause of the fire will most often be carelessness or negligence.

Houses of Worship

In the case of a fire in church or other building of worship, it will be necessary for you to make a most complete investigation and make a satisfactory report—the public will demand it. Like any other fire, one in a house of worship is almost invariably *accidental*. After determining that there is no reason to suspect arson, you can proceed in the same manner as with any other fire investigation, except you will have to consider objects associated with the particular religious services held in the building, such as an organ, candles, etc., as being possible causes. Due to the complicated architecture of most buildings of this type, any delay at all in discovering the fire, will almost assure total burning. The investigator should begin by questioning any one familiar with the building. With some knowledge of the layout of the building before the fire, the investigator can then enter the ruins and reasonably determine the path the fire followed.

The investigator must use extreme caution before entering a church, because the huge rafters supporting the high, peaked roof almost always collapse into the building. Since these rafters are precariously situated to begin with, it is easy to understand that the slightest weakening of the wall, or burning of the ends of the rafters, will cause them to come tumbling down. Many churches hold social events in their basement halls or even rent them out. If the fire appears to have started in the basement, ascertain what it was used for last or if it is just used for storage. If an event was held in the basement even a day before the fire, it should not be too difficult to locate the last persons to leave and question them. They will be able to tell you whether or not the cooking equipment was used and if it may have been left on. They may also know whether the hall had been cleaned up after the affair and the trash cans emptied outside the building. The lights, fans, heater and other equipment which are left on indefinitely may not normally cause a fire in a church basement, but if there was an affair where decorations and crepe paper was strewn about, the fire could have been started by a light bulb.

Since it is reasonable to assume that no one smokes in the part of a church where the services are held we can discount this possibility if that is where the fire originated. However, faulty or overloaded wiring should be given prime consideration. In most older churches the original circuits are inadequate for the new floodlights on the parking lot, the lamps along the sidewalk, or other additions made since the church was built.

Rockets

With all the publicity given to the space program and rockets, a breed of amateur experimenters has developed. Many of these, mostly young experimenters, have been responsible for fires, explosions, crippling and even death. This usually results from the junior engineer not being totally familiar with the chemicals he is mixing for fuel to power his rocket. The most common type of rocket made by youngsters is a piece of pipe closed at one end and crammed with the heads of matches in the other end. This experimental work is usually done in the cellar or garage of the family home. The trouble does not start when the rocket is launched in an open field, as intended, but when it goes off prematurely in the enclosure—the most likely cause being friction. When the investigator arrives on the scene he will probably find the empty chemical containers, match sticks without the heads and parts of the rocket. In the case where chemicals were purchased for the experiment, he may find the bill of sale for them and even well-drawn plans for the rocket, telling how the chemicals were to be used. With this information and the help of a chemist he should be able to tell exactly what caused the rocket to fire prematurely.

SUMMARY

1. Unless arson is prevalent, or you know an arsonist is working in your area, most of the fires you investigate will probably be *accidental* in origin.

2. Small children are often responsible for *accidental* fires and information to this effect will not always be readily volunteered by the parents. The parents may even deny the child's involvement before you suggest it.

3. Make a hobby of studying and classifying personalities and people's reactions at a fire and while being questioned. This knowledge will be an asset when you approach people and will give them confidence in you instead of fear of prosecution.

4. The clue you need to solve the question of the origin of an *accidental* fire may be found in an unburned portion of a building. Be observant for cigarette burns on ledges, overloaded wiring and antiquated or faulty appliances. If these conditions exist it is safe to assume they were prevalent in the area where the fire started.

5. Elderly or feeble persons, children left alone in a house, and drunks unfortunately have the same disadvantage in common: They are unable to comprehend danger.

6. If you arrive to investigate a fire and find only the chimney and fireplace standing, it is not an indication that they were not involved in the origin of the fire.

7. Heating units are often the cause of ignition in fires. This is not necessarily due to the unit being faulty, but to placing inflammables in too close proximity to the heat. This is especially true of portable units.

8. Public opinion, as well as insurance companies, will demand a complete report and thorough investigation where a fire involves a house of worship. Leave no stone unturned in making this type of report satisfactory.

Chapter Four

PYROMANIACS
METHOD OF OPERATION

THE fires set by a pyromaniac are similar to fires set by juvenile fire-setters: they are generally small and hastily set. Investigation will show a lack of motive. Generally speaking, the pyromaniac will use matches to start the fire rather than use a time device. He will burn paper or trash which was already at the scene, rather than use gasoline (unless it is readily available). He will set fire outside of a building, under a porch, in alleys, in hallways or any place accessible to the general public.

A series of fires of this nature, which have been set in a similar manner, indicate, a pyromaniac at work. This type of person likes to do his work after darkness. As time goes by he will set more serious fires and set them more often. Many pyromaniacs begin by setting field fires, then vacant building fires, and then building fires of any nature. He will give no thought to whether or not people are in the building, or if there is danger to them; his only interest is the fire itself.

INVESTIGATION

When the facts show there is a pyromaniac at work in your area, you should make a check of all known persons of this type living in the area. If the fires show a pattern, it may be possible to stake out the area where the next fire is expected. The investigator should arrive at the scene of the fire as soon as possible after the alarm is given. On responding to the fire he should note and check on all autos and persons leaving the area of the fire. The pyromaniac will set the fire, leave the scene, and then return to the scene after the fire engines have arrived and a crowd has started to gather. The crowd should be watched as the investigator may recognize a known pyromaniac in the crowd. He may appear excited or in a daze at the scene of the fire. His eyes may appear glassy or he may have an excited type of stare. Many photographs should be taken at the fire and these should include the crowd. This can be done by using the crowd as a background for pictures of the fire. These photos of the crowd, when examined, may reveal a known pyromaniac in the crowd. Also, these photos can be compared to those of other fires in question and may reveal unknown subjects who are present at all of the fires. These subjects can then be investigated.

If you have no known pyromaniacs in your area, or if they all check out as not being involved, then you should check on your juvenile fire-setters of the past years—one of them may be your pyromaniac of today.

A check should be made on any reports of larceny of women's clothing in the area, as this is sometimes tied in with pyromania. Prowlers and "peeping Toms" should also be investigated.

Often a pyromaniac will actually reach an orgasm while setting or watching the fire. Due to their nature, there have been cases where pyromaniacs have been known to direct their fires against happy and gay people. He need not know these people personally, and they may not know him at all.

You, as the investigator, will question anyone you think may give you some information. While you are doing this you may notice someone in the crowd changing places to avoid being questioned and moving around to mingle with the persons you have already questioned. If this person is the pyromaniac and he has already had his satisfaction, he may be curious to know how much you know and begin a conversation with those persons you talked to. On the other hand, he may even approach you and offer some irrelevant information to throw you off your investigation or hoping you will discuss what you know about the fire. If his mind is really working fast, he will invent a suspect for you and may describe a vehicle he saw leaving the scene. This can also be applied to juvenile fire-setters when they think you are getting close to finding they are responsible for setting a fire. When you think you are dealing with a pyromaniac, re-member his type of personality: before the fire is set he is careless and anxious and only wants to satisfy his desire; after, he has accomplished his desire, he may be scared or repentent. What-ever his emotions after he has satisfied himself he is vulnerable to an alert investigator and if you can put your finger on him at this time you will never have a better chance to close the case.

QUESTIONING SUSPECTS AND
THEIR RELATIVES

If various phases of your investigation suggest a particular person as being a pyromaniac but questioning is to no avail, you may find the relatives of the suspect more helpful if prop-erly approached. These people realize there is a problem and if you call them to your office or station to discuss the problem, impressing on them the position they are in and gaining their confidence, you will find they readily cooperate because they have long been hoping for someone who can help them. A long history of the suspect's relatives will then unfold. Cases of pe-culiar behavior, isolationism and possibly some sex oddities will be brought to light. During this conversation the investigator should listen tentatively and sympathetically and soon the suspect will be brought to you with orders from the relatives to tell you everything. Close relatives of a pyromaniac may, at first deny your suspicions. But when they are convinced you are the one to help them, and there is a suggestion that you already have some incriminating information, they will readily cooperate. A close relative of a pyromaniac who has lived in proximity with the subject will have noticed some peculiarity about which he would like to tell someone who can help. However, the relative may need a little prompting. A few inquiring suggestions about the personal habits of the suspect will ease the embarrassment of the relative if he is reluctant or does not know where to start. If you are first to mention a peculiarity, relatives will feel free to talk about it. We are not sure whether the psychological reason for the cooperation of relatives, in certain cases, is due to their genuine desire to help the suspect, or if, after years of tolerating the shame, they are thankful you came along to unburden them. We do know, however, that once we gain the confidence of a suspect's relatives, they will be of great help to us. Accept whatever reason they give for cooperating, because it will probably be something *you* suggested, such as civic duty, protecting the innocent or helping the suspect, instead of *the real reason* for exposing their secret.

It is best to talk with the suspect separately, rather than with the relatives present. A suspect who has a problem will rarely give you any information in front of a member of his family who will make him feel ill at ease, or continually interrupt and reprimand while the questioning is in progress. As a matter of fact, he would prefer his relatives not be informed at all. The relatives would not be able to talk freely in front of the suspect anymore than you would talk about a friend if he were standing right next to you. The relatives would feel guilty and would rather have the suspect think the investigator was the villain and responsible for the apprehension. If it will help to close our case, we, as investigators, will accept all the blame for the apprehension and even agree with the relatives when they say the authorities are responsible because they did nothing about the suspect the first time he was in trouble. When discussing relatives of pyromaniacs, we include husbands or wives, as well as parents or guardians. The mate of a suspect would be an excellent source of information if you could get him (or her) to discuss their married life, but for obvious reasons this would take much diplomacy. In some cases, neighbors of a suspected pyromaniac will give you more information about the success of his married life and his habits than can be obtained from his wife. Adult pyromaniacs will sometimes live with their parents, guardian or a relative even if they are married. Married life may not meet expectation, or the demands and responsibilities of married life will not coincide with the private little world the pyromaniac recoils to when the urge overtakes him. Any persons who have been involved with the pyromaniac can be excellent sources of information if you can get them to confide in you.

Case History No. 1

A large number of vacant buildings in the area were burning. These included sheds, barns and vacant houses. All were located in places where there were no houses close by, or where the view from another

FIGURE 17. This barn, of frame construction, is burning quite well, as you can see. A look at the smoke in the picture will tell you that the wind is blowing to the right of the picture. The fire seems to be burning best in the top center of the roof. This is because the small construction on the roof at that point is a ventilator. With the doors and other openings around the bottom of the barn and this ventilator open at the top, it causes a perfect draft to carry the flames right through the barn and up to the roof. The firemen having their helmets on backwards is an indication that this fire is giving off a great amount of heat. They are using the long rear brim of their helmets to protect their faces from the heat and sparks.

house was obstructed by trees or other objects. Because of the lapse in time in reporting most of these fires, the damage was extensive, if not complete. An extensive search of the area around each fire revealed that no one had been seen entering or leaving the area. The investigators, in each case, found a two-gallon can either in the ruins of the fire or in nearby bushes. A laboratory examination revealed that the cans had contained gasoline. In some cases, the investigator was able to recover, from buildings involved, several pieces of wood which when examined, showed signs that they had been drenched with gasoline.

The investigators then reviewed their records of past fire setters in the area but the results were negative. The investigators then got together with the police officers who worked the area. With these men who were very familiar with the area, they were able to pinpoint the other vacant buildings in the area. They found that they had more vacant buildings than they had men. The closest neighbors to each vacant building were contacted and their aid in watching the buildings was enlisted. The police patrol cars were requested to keep a close check for suspicious people walking or autos parked in the areas around the vacant buildings. Several nights later there was another fire. This vacant house was located on a dirt lane and was in full blaze before the fire department was notified. The investigator and several police cars arrived on the scene right away but found nothing. For a while, no one could get near the house—it was in full blaze and the police were busy keeping the dirt lane clear of sight-seers and holding the crowd back that was forming at the scene of the fire.

During this period the investigator took many pictures which included the crowd that had gathered. He questioned as many people in the crowd as was possible. All investigation at the scene appeared to be negative. But upon questioning the people who lived in the house where the dirt lane met the paved road, it was found that a woman, shortly before noticing the fire, had heard her dog barking; but when she looked out she saw nothing. The investigator then questioned her further and found that early in the day she had seen a car go back the lane, but it came right back out. She said she thought nothing of it because cars often go back there. When asked for a description she could only state that it was a late-model shiny green auto driven by a well-dressed man about thirty years old. She said she did not see this auto in the area at the time of the fire and saw no one leave the area on foot.

This information was given to the police in the area and one of the patrol cars replied that he had observed a green, late-model auto parked several blocks away just before the fire. This officer had noted the make, model, year, color and tag number of the auto. A check through the department of motor vehicles revealed the auto was listed to a business. A check with the business showed that this auto was used by one of their salesman who was thirty-two years old. This subject would have no reason to be in that area at that time of the night for his business.

All cars working the area were advised of the above information. A few days later, this auto was seen leaving a vacant building in the area. When a close check of this building was made, it was found that there was a two-gallon can filled with gasoline hidden near the building. This building was then staked out. That night, the subject returned to the building on foot and removed the can of gasoline from its hiding place. He than began pouring the gasoline in the vacant building. He was arrested as he was about to light the fire. His auto was found parked several blocks away. Upon questioning, he stated that during the day, when he was working, he would locate the building he was going to burn and hide a can of gasoline at the scene. Then he would return after dark, park his auto several blocks away and walk to the scene and start the fire. He would then hide in the bushes or somewhere else on the scene until a crowd started to gather. Then he would slip out of his hiding place and join the crowd in watching the fire. Once caught, he admitted to setting all of the fires and admitted that he needed help. The only reason he could give for setting the fires was that he had an uncontrollable urge to do so.

This subject was found guilty of his crimes and placed in an institution where he will receive treatment.

Case History No. 2

This case began on a Wednesday night around 9:30 PM, when a small fire in the laundry room in the basement of an apartment building was discovered. It was found that several pieces of clothing had been piled in a corner and set afire. The fire did very little damage, but caused a lot of excitement as the tenants rushed out of the building. Normal investigation at this time proved negative. On the following Wednesday night around 11:00 PM there was another fire in a laundry room in another of the apartment buildings. In this fire some clothes that were left on the scene and some newspapers were burned. Again, no real damage was done. The following Wednesday there was a fire at 9:45 PM in the basement of another of the apartment buildings. This building, instead of having a laundry in the basement, had storage bins for the tenants to store their surplus goods in. This fire was in a storage bin and did extensive damage to the personal goods stored there and threatened damage to the whole building.

During the next several weeks, there was a fire in one of the apartment buildings every Wednesday, except one, between 8:00 PM and 11:30 PM. Due to the large number of people coming and going from this apartment building during this span of time, it was difficult for the investigators to determine who was suspicious during their attempted surveilance of the area. At this point the investigators had no real description of the subject wanted. The one thing in the investigators' favor was the fact that the subject only set fires on Wednesday between 8:00 PM and 11:30 PM. They had already determined that there were thirty-three apartment buildings in this apartment area. There were too many lives at stake to play the hit-and-miss game any longer; the investigators requested help and got it. On the following Wednesday there was a plainclothes law-enforcement officer in the basement of every apartment building in the area. After a long wait at approximately 10:55 PM a police officer arrested the subject as he was about to light a fire in a storage bin in the basement of one of the apartments.

The subject arrested was a middle-aged man of average income. After extensive questioning he admitted that he had set all of the above fires. When asked why he had set the fires, he could give no reason. It was found that the reason he only set fires on Wednesday was this was the only night he could get out alone. His wife kept a very close reign on him but on Wednesday nights she went out with the girls. She would usually leave home around 7:00 PM and return home around midnight. The Wednesday that there was no fire, she had been sick and stayed home.

This subject was found guilty and sentenced to the penitentiary where he will receive treatment for his illness.

SUMMARY

1. The pyromaniac generally will hastily set small fires with no apparent motive.
2. Pyromania can be described as a progressive illness.
3. Records should be kept on all known pyromaniacs living in your area.
4. Photographs should be taken at the scene of the fire and should include the spectators, since the pyromaniac will often stay to watch the fire.
5. The pyromaniac may keep changing places in the crowd in order to keep from being questioned as you make your rounds.
6. The relatives of a suspect may be more than helpful if they feel that you truly want to help the subject with his problem, not hurt him.
7. When questioning a suspect, in most cases, you will be more successful if you question him alone rather than in front of relatives or friends. He would rather bare his secret to you than to relatives or someone who knows him.
8. Neighbors of the pyromaniac can often give you very interesting information as to his past and habits.
9. Remember the pyromaniac is a sick person.

Chapter Five

ARSON

WHEN the investigator fails to locate a reasonable cause of accidental origin of a fire, or is unable to get to the origin at all, or if he uncovers any suspicious circumstances, he should immediately change his routine investigation to an arson investigation. An arson investigation is a criminal investigation and should be conducted as such.

ARSON INVESTIGATION PROCEDURE

Steps should be taken to rope off the area and to do everything possible to protect the crime scene. Announce your suspicions to the ranking fire official at the scene; you will need his assistance and he will be glad to cooperate and assist you. Firemen, in most areas, have the noble habit of cleaning up the debris after a fire by shoveling it out the window into the yard or alley. This may prevent recurrence of the fire and may be good public relations, but to the arson investigator it is a handicap. Request of the fire official that his men not disturb the crime scene any more than necessary. It is easier for you to examine the ruins while still at or near their location before the fire, than to sift through the rubbish piled in the yard trying to determine what part of the building it came from. You may now proceed with your arson investigation.

Other Crimes Involved

The motive for the arson may be readily apparent. It may be to cover up another crime such as burglary or murder. Burglary can be suspected when the first firemen to arrive relate that the door was already open on a building where no persons were au-

FIGURE 18. In the front of this picture we see a mattress that has burned. In the background to the left of the open door we see the bed itself. We are looking at the headboard which is burned away. The headboard had been covered with a clothlike material. This added to the fire. The room with the open door is where the bed was previously located; as you can see through the door and window, very little damage was done to the room. It is difficult to tell in the picture because of the mattress having been pulled apart by the firemen to put the fire out, but the fire did burn mostly on the head-end of the mattress. The damage to the headboard also agrees with this. The cause of the fire was found to be "smoking in bed."

thorized to be inside at the time of the fire. Examine the doors and windows of the building for signs of forced entry. If the building was ransacked, or valuables known to have been in the building are missing, all steps taken in a criminal investigation should now be used. Photographs should be taken as early as possible of the burned as well as the unburned portions of the building. Pictures of all rooms in the building, doorways, windows and a general view of the fire may be helpful later in the investigation; you will often see interesting things in photos that you failed to observe at the scene. Also, it is easier to concentrate on a section of a photograph than to observe an area with the

FIGURE 19. A sign of another crime is shown here. Forced entry had been made by breaking out the lower right pane of glass shown in this picture. The night bolt is a very good idea but it was placed too close to the window. The subject just reached in and unlatched the bolt. The curtain was pushed aside so the photo could be taken. The fire was in another part of the building and if a complete fire-scene search had not been made, this point of forced entry may not have been found for some time. In this case, the fire was set to cover up the first crime or at least to destroy fingerprints and other evidence.

distractions of the activity going on around you. Among the pictures you take, be sure to include shots of pry marks around doors and windows, open record-books, electric clocks and close-ups of contents of the building that may have been scattered and meant to burn. Particularly cover well the particular area where you think the fire began.

Arson and Homicide

In cases where arson and homicide are involved, the value of pictures cannot be overemphasized. If the fire is still in progress

FIGURE 20. Two young children died on the floor shown here next to the bed. At the top of this picture you can just see the bottom edge of a window sill. They didn't quite make it. This bedroom was located on the second floor of a frame house. This house was engulfed in smoke and flame when the first police and fireman arrived on the scene. As you can see actually this room was not reached by the flames, but the heat and smoke was very great. The bed clothes in the picture are darkened and charred from the heat but not really burned. The walls and furniture are blackened with the smoke and soot. A rubber covering that was on the steps leading upstairs and in the hallway, no doubt, added to the heat and smoke. The fact that the children were found on the floor and that some of the bed clothes were pulled to the floor shows that the children had been alive and attempted to get out.

the body will have to be removed immediately, as fire may destroy any or all clues to the arson or the homicide. Try to take as many pictures as possible before the body is removed. Also, measure and identify as many objects in the room as possible; and measure the location of the body in relation to some key objects in the room. These steps will be invaluable to the homicide investigators. Smoke and heat are hard on fingerprints and other clues, so get whatever cooperation from the fire department you can in reference to this. Most fire departments have fans and knowledge of ventilation procedures that will rapidly air out and cool off an area quickly when necessary. Fire investigators should be quick to notify the respective divisions of their department when arson coupled with another crime is suspected, because other specializing investigators will want an early, on-the-scene investigation.

Fire Setters

Arson is a broad field and can cover many areas and have many motives, sometimes even without material motive, as in the case of pyromaniacs. Arson by a professional fire setter, known as a "torch," is usually done for insurance purposes or some other monetary gain. The "torch" enjoys his work and is paid well for one night's (or day's) work. He will often go through elaborate preparations to make his fire as complete and spectacular as he can. This will entitle him to his pay and maybe a little bonus. The fire set by the "torch" will not be confused with that set by the pyromaniac, or by the candle-burning citizen, or by the business man who could not afford or could not contact a professional torch. A "torch"-set fire can be recognized by its completeness; the investigator may have to turn to examining the financial status of the occupants in order to determine who would gain by the fire.

QUESTIONING PRINCIPALS

Remember, when questioning persons connected with a suspected arson, these people are businessmen and good citizens; they are not experienced criminals and can only endure a certain amount of questioning by the professional investigator. Unrelenting cross-examination of a suspect, in unfamiliar surroundings such as an interrogation room of the police station, with a folder of evidence on the table in front of him, will often shake the good-citizen arsonist to a fatal slip of the tongue. A good investigator can detect this slip and repeat it to the suspect in a manner that will cause the already tense and unnerved citizen to confess. As soon as this happens and before he recovers his composure, get as much information and as many names as possible; then get the statement blanks and a witness or stenographer. Of course, the motto of known criminals who have been had before will be "Don't say anything." In this case, it may be helpful to let the suspect think one of his confederates has already been picked up and has implicated him. None of them wants to "take the rap" alone. Impress on the suspect how happy the judge will be when the investigator tells him the suspect cooperated with the authorities.

WHEN TO SUSPECT ARSON

There are many circumstances that can lead a trained investigator to suspect arson. There may be empty containers at the scene which contained accelerants, or burnt matches, empty matchbooks and flammable materials in the area where the fire started which are not normally there. Evidence of a trail, usually along the floor, is a sure sign to consider arson. The trail can be best described as a fuse and is used by an arsonist to help spread the fire or make it appear the fire started in a different area than it did. Usually the only evidence of a trail will be a narrow burned area between two other burned areas. You should have some of the residue from the trail sent to the laboratory to find out what type material it was. If witnesses report persons or vehicles leaving the area just before a fire, or suspicious persons in the area even days before a fire, or prowlers around neighboring garages or storage sheds, arson should be considered. Other suspicious circumstances could include establishments that have a watchman who was off sick the night the fire started, or have faulty automatic fire alarms and sprinkler systems, or empty fire extinguishers. While some of these conditions may not have been

FIGURE 21. This is a piece of furniture that was in a home that burned. The finish on this piece of furniture has been badly damaged by smoke and heat. The walls around this furniture were covered with wallpaper, which you can see is burned and peeling. One thing in this picture that may raise question for the investigator is the reason for the ice-cube trays being on this piece of furniture. Anything that is found in a place that is other than normal for it should be questioned by the investigator.

deliberate, they can be considered negligence and will be of interest to insurance companies. Any condition that either hampered extinguishing the fire or seemed to accelerate the progress of the fire should be considered by the investigator to be a suspicious circumstance and should be thoroughly investigated. A professional arsonist will know that a fire needs the proper draft to get started and spread in the direction he wants it to and may open doors and windows to facilitate this. An investigator should be especially wary when finding outside windows open in the cold season when they normally would not have been left open. Windows on the ground floor of a commercial building are rarely left open after closing in any weather, if they are accessible from the ground; but an arsonist would consider these windows the best to open to create the proper draft for total burning.

TIMING DEVICES

Many arsonists, both experienced and amateur, consider a timing device a *must* when setting a fire. The timing device, as you know, permits the arsonist to establish an alibi by being in a different location when the fire starts, possibly with some friends who may be totally innocent and can swear to his presence. There are timing devices too numerous to mention that range from a slow-burning candle to elaborate electronic devices connected to timers that will permit the arsonist to be hundreds of miles away when the fire starts. Cigarettes, springs, wires, clocks or parts thereof, or any unidentifiable mechanism found in the ruins of a fire can indicate a timing device but would most likely be used as such when the arsonist wants to set off a highly flammable or explosive substance and just wants enough time for his own safety. A fuse will usually burn too fast for the arsonist to get to another location and establish an alibi.

POSSIBLE SUSPECTS

All arsonists, however, do not use timing devices nor do they all prefer to have an alibi placing them at another location at the time the fire started. As a matter of fact, you may be overlooking your best suspect by not considering the hero of a fire, or even a victim who was burned or injured in the fire. In suspicious fires, relatives of victims who appear to be grief laden cannot be overlooked. Of course, this situation must be handled delicately and with much diplomacy, at least until you find some proof of your suspicions. It would be human nature for a novice arsonist to think he would be the last one suspected if he were to receive some minor burns while attempting to save some of his property from a fire. This same reasoning should be considered in suspected arson, where a surviving relative fails to save a victim who, in your opinion, could have been saved in the early stages of the fire. This suspect will often loudly proclaim his criticism of the fire department, which was not notified until the house was engulfed, for not effecting a rescue that he himself had the best opportunity to do when the fire was in the early stages. A novice arsonist may even put on a pretty good exhibition of trying to put out the fire and then upon arrival of the fire department try to be of assistance to them. This exhibition is for the benefit of anyone concerned so they will know how anxious he is to put the fire out and know that he could not have possibly started it.

HOUSES OF WORSHIP

The house-of-worship fire can be a very difficult and interesting fire to investigate. Many houses of worship are large in size and have very large rooms with high ceilings, all of which can make for good and fast development of a fire that has started. Churches are often old buildings and the electrical wiring is also old and outdated. In some cases, because of money problems or whatever other reasons there may be, needed repairs are not made. Consideration must also be given to the fact that there are drapes, choir robes, and many other cloth items as well as candles, which are kept in many churches.

One of the first steps in investigating a church fire should be to talk with the reverend of the church and also with the caretaker. Find out if there was a service or any type of meeting being held in the building at the time the fire was discovered or shortly before. If there was no meeting in the church, ascertain whether anyone was in the church prior to the fire, whether

the doors to the church normally are kept locked, and whether
they were locked when the fire was discovered. Find out who
has keys to the building. Ask the reverend if candles are ever
used in the services and if they are ever lit. Many churches that
do use candles seldom, if ever, light them. You will want to
know where the candles are stored and whether any candles
are left out in the church, such as around the altar. It is important to know just where any candles that are left out in the
church should be located and if possible just how many candles
should be at what location. Any candles found at a location where
there should be none would be a sure sign of arson. If the building is usually kept locked, then we should look for signs of
forced entry. A check should also be made to see if the fire was
set to cover up another crime or accidentally set by someone
while committing another crime. Many churches have safes which
someone could have entered the building to crack. Of course,
as pointed out in other chapters, many church fires will be set
by women or juveniles, in which cases, it will not usually be to
cover up other crimes. And, of course, the fact that many churches
are left unlocked, is an open invitation to any pyromaniacs who
may be in the area. There is also the chance that if the church
was unlocked, children playing in the church may have accidentally set the fire.

When talking of church fires we must also keep in mind the
hate groups that often commit this type of crime. Often these
people will use some type of explosive in their work. So if an
explosion has been heard and the boiler has not blown and no
one has blown the safe then there is a good chance that you are
up against the work of a hate group. Of course, these groups
will also use the conventional fire-setting ways to do their work.

As you can see there are many people who can be suspect when
there is a church fire. But if there was a group of people in the
church when the fire started then you can gain much information by questioning them as to when they first saw the fire,
where they first saw it, and how it started. If there supposedly was
no one in the church, and no candles were left burning in the
church, then there is little reason for a fire to start without human help. Therefore, anything odd or out of place should be
investigated very carefully. Did the fire start in an odd place?

Did it burn faster than normal? Was the smoke an odd color or smell? Talk with the fire fighters who were first on the scene, they can be a big help with these questions. Always check the area around the fire scene for containers that may have held flammable liquids. This will sometimes be dropped at or very near the scene of the fire. Check any bushes or wooded areas where these may have been thrown. If any items were stolen from the building there is also a chance that all or some of these may be found hidden nearby. There may even be things dropped by the criminals that they did not mean to drop. It is not unheard of that a subject may drop his wallet or other identification at the scene of a crime.

Cases at Random

It is not always necessary to conduct an intensive investigation to solve an arson fire. In some cases, the fire setter has no fear of apprehension or may even want to be caught to further emphasize or draw attention to his reasoning. There was one irate husband who so objected to his wife's betting on the horses and losing money that he vented his spite on her by setting fire to their home. This man was so emotional about his cause that he even threatened the firemen who arrived to put out the fire by telling them he would shoot them if they tried to stop him. This case bears out the well-known fact among law men that emotions run high in domestic cases and this adds to the difficulty in handling them.

In a suspected arson case on the West Coast an autopsy on the victims indicated multiple crimes including murder, arson and suicide. The perpetrator, in this case, was found nearby lying on the shotgun with which he had killed himself. He was related to the victims and this proved to be the climax of another one of those highly emotional domestic cases.

A nine-year-old boy who had the presence of mind to roll on a mattress after being set afire by an older boy, probably did much to save his own life. The older boy, age fifteen, became irate when the younger boy refused to run away from home with him. Some paint thinner as an accelerent helped turn the younger boy into a human torch.

Figure 22. This picture shows an extensively damaged home that was the result of arson. During the night, while the family was in the home, some unknown person (s) threw a Molotov cocktail into the home. The family was very lucky and escaped without injury. But as you can see the damage to the home was great. This type of fire-setting device is becoming more and more popular. This device rapidly causes a very hot fire.

A young arsonist apparently bent on killing an entire family of seven on the East Coast threw Molotov cocktails through the living room window. When the victims awoke from the sound of the breaking glass they looked out the window in time to see the arsonist still standing in the yard. This was about 4:00 A.M. and neighbors reported seeing a suspicious person around the house during much of the night but did not think to call the police. The suspect had such a strong grudge against one or more members of this family that he even slashed the four tires of the family car that was parked out front. Had not the victims made their escape through the flaming living room, this would have been one of the most serious cases of arson and homicide on record.

FIGURE 23. In this photo we are looking at the floor inside of the doorway of a home that was the target of a Molotov cocktail. As you can see, there is a large amount of debris on the floor which was the result of both the fire itself and the firefighters doing their work. This is the type of condition in which the fire investigator will be required to work. In this case, it would be necessary for the investigator to sift and sort through this mess in an attempt to locate pieces of glass from the bottle that was used in making of the cocktail or any other evidence that may be found at the scene.

SUMMARY

1. When arson is suspected, rope off the area of the fire and treat it as a crime scene.
2. If another crime was involved, you will be able to establish the motive for the arson.
3. Take photographs, even while the fire is still in progress, and if homicide is suspected, do this before removing the body. Take measurements in relation to key objects in the area where the body is found.
4. In cases of burglary and arson, identify the "M. O." of the burglar. This will help to classify the burglar. A thief will often take items he has taken before and found he can unload, such as furs or diamonds, while others prefer just to take cash. Some have an affinity for small objects that they take for their own use, such as transistor radios or phonograph records. Noting these facts will help tie in an offense with other cases.
5. Look for remnants or residue from timing devices. You may have to get on your knees to sniff the burned floor boards. Look on the floors below where the fire started for signs of an accelerant having leaked through.
6. If you find it necessary to remove a piece of flooring, molding or other object that is secured to the building, it may be well to make this decision while the firemen are still at the scene with their handy tools and experience in wrecking. You can then take the object with you when you leave and have the lab get an early start on its analysis.
7. Learn how to question people by first characterizing them. There is the saying in police circles that "more flies are caught with honey than vinegar."
8. Arson involved in house-of-worship fires is increasing in various areas. This is more likely due to the house of worship being used for nonreligious gathering than any plot against religion.

Chapter Six

OTHER TYPES OF ARSONISTS
THE JUVENILE FIRE-SETTER

THE fire set by a juvenile will generally show lack of expertness. The favorite method of starting the fire is by using matches to light paper or other burnable materials that are available.

FIGURE 24. This picture shows some burned paper on the floor in a vacant house. This was a case where more than one fire was set. The fire department was called to combat a rather large fire in another part of the house. In this room paper was piled on the floor and lit, but it just burned out. Matches were found on the floor around this area. The damaged wall, shown in the upper part of the picture, has no real connection with the fire other than it was one of the many signs found, that showed children had been playing in the house. Questioning of the children who were watching the fire proved very helpful in this case. The same juveniles who were responsible for this fire were found to be responsible for several woods fires in the area.

Although the above is usually true, it must be kept in mind that juveniles do use other methods. Recently, the authors investigated a case where an eleven-year-old boy attempted to set fire to a barn by piling wood, rope and other debris, and then dumping kerosene on the pile and using a candle as a timer. Working on information that he was in the area of the barn at the time of the fire, the boy was questioned and he soon admitted that he started the fire. This shows that although the timing device is not considered to be part of a juvenile fire-setter's method of operation, the fact that one has been used cannot in itself rule out the juvenile.

The fires set by juveniles will usually be set during the non-school daylight hours or shortly after darkness. Field and woods fires are very often set by the juvenile fire-setter. Often the juvenile will set fires in spots near existing fires such as near an incinerator. In this case, they will light paper from the fire in the incinerator and use this to start their own fire nearby. One reason for this may be the fact that they cannot get matches of their own. In most cases, these fires will cause minor damage or destruction.

We had another case where a male juvenile using matches and existing debris set fire to the barn next to his home. The barn was large and there were several homes very close to it. Early discovery by neighbors and prompt attention by the fire department prevented a major fire. After investigating the fire scene and questioning the neighbors, it still took some very serious questioning of the juvenile before he would give a statement. Even then he could not or would not give a reason for setting the fire.

Barns and vacant buildings are popular places for juveniles to set fires. Often they will set fires under houses. If a juvenile sets a fire in a house, many times it will be in a closet, or he may set a bed on fire. In one case, a five-year-old boy set his bed on fire and thus caused his whole home to be gutted by fire. Several of his younger brothers and sisters were in the room with him when he set the fire. Fortunately, none was injured. When questioned he stated, "Yes, I burned my house down." When the mother was questioned she stated that several months before

FIGURE 25. Here we are looking at the rear wall of an outside storage shed. The shed was approximately 10 ft. X 10 ft. in size of frame construction. As you can see, the building was just a few inches off the ground. The fire started just under the right-hand corner shown in the picture. It then worked upward and outward. The lower few feet of the building on the left side of the picture did not burn too much. This was due to the natural tendency of the fire to burn upward plus the fact that concrete blocks were piled against the lower part of the wall in this area. This fire was started by pre-school-age children playing in the area.

he had set a bed afire but she found the fire when it was still small and put it out herself. The fire was never reported.

We must remember that the juvenile fire-setter of today may be the adult fire-setter of tomorrow. Therefore, it is important that these fires be investigated and the offender apprehended. A good record of all juvenile fire-setters should be kept for future reference. This is pointed out in our chapter on pyromaniacs.

Question All Juveniles

If an investigator fails to find an indication that the fire was started accidentally, and his investigation then goes to a pyromaniac or juvenile, the juvenile would be the best suspect unless there was a pyromaniac known to be working in the area. If the area where the fire started was accessible but hidden, the percentages are that a juvenile was involved. This will often, but not always, be a vacant or abandoned building or shed that, in the juvenile's mind, no one wanted anyway. During and after the fire it is not likely that any juveniles will leave the scene. Especially the younger ones, to which we are now referring, will remain at the scene until the last piece of equipment has left. This is a good time to talk to them and find out who is missing from this spectacle, and who was seen running from the area of the fire, even if a couple blocks away, while the rest were running toward it. Be especially interested in any juveniles who were seen in the area at any time during the day or who normally play in the area. It is not natural for this child to be missing from the area now unless he is out of town. The sirens and red lights will draw the children to the scene no matter what they are doing and those who do not put in their appearance are good suspects. It is also possible that the fire setter was at the scene and decided to leave when he saw you begin asking questions. As we said, it is not natural for a young juvenile to leave the scene while there is still activity going on.

When questioning the children, be sure to question the girls as well as the boys as it is normal for them, especially in the very young, to want to tattle on each other. Sometimes a small child playing with matches will start a fire deliberately in some place where no one can see him because he is not supposed to be playing with matches and the fire will accidentally get out of control. This child will almost certainly leave the scene in panic and may even have burned his hands while trying to control the fire. At any rate, any of his personal belongings or toys he had with him will be left at the scene. While it is difficult to believe, we know juvenile fire-setters to, at times, be very young—even "toddlers." Of course, the only intent of those children is to imitate something they saw an adult doing. We do want their parents informed of the dangerous habits the children have acquired, even though no charges are made.

FEMALE FIRE-SETTERS

The fire set by a female fire setter will often resemble that set by the juvenile fire setter, as it will show a lack of expertness and may not make too much sense. The female will generally set her fires inside the home, usually in the kitchen, in closets or in the wardrobe. Often her fires will be meant to mar or deface but not to destroy. Matches will be used to start the fire and, if an accelerant is used, it will be a cleaning fluid or something else which can be found in the home. There will often be small and numerous fires. Fabric items, such as curtains, draperies, linens and clothing, are most often burned. These items may be the property of the fire setter, her husband or her boy friend. She will often set fire to her own personal property or things of sentimental value to her. Many times a female will just turn the cooking stove on and throw the items to be burned on top of it.

Type of Woman

The female fire-setter is a nervous, angry or upset woman. She wants new drapes or clothing and her husband will not buy them, so she burns the old ones so he will have to buy new ones. If she has a fight with her boy friend she may get mad and burn his things. At this time she may also burn things of sentimental value. This type of woman does not need a great reason to do her burning. Her husband may fail to hang up some of his clothes so she will burn them up. The female who is "fed up" with housework is the type who will toss the dish towels, and etc. on the cook stove and turn it on.

As you can see most of these fires are set by a woman who is mad at her man. This is, in many cases, an easy fire to investigate. Often, on your arrival the woman will readily admit she set the fire. She may even state that she fixed that lousy no-good cheap so and so. If not, questioning of the boy friend or the husband, as the case may be, will no doubt shed some light on the case. In either case, proper questioning of the female by a good investigator should result in a confession, in a very short time. During questioning of the female, it is suggested that you have another woman present.

FIGURE 26. This fire occurred in the kitchen of a modern home. It was arson. The fire started at the kitchen stove and did extensive damage above this area. The rest of the house received only smoke and heat damage. This fire was started by placing clothes on the stove and then turning on the burners. Between the stove and the sink you will see a cloth item hanging off the counter. By looking closely you can see that these are the sleeves of coats. You can see the sleeve buttons. This, of course, is not the normal place for coats. The clock in the built-in oven will give you the time that the fire caused the electricity to shut off. This will help you determine the time the fire was set.

This same type of woman has been known to set church fires. When she sets a fire in the church it will generally be in the cloak room or where the robes are stored. It might be interesting to note that, there again, she is burning fabrics.

ATTENTION-SEEKING FIRE-SETTERS

This type of fire-setter will set a fire near his home, his girl friend's home or near some place over which he has charge. He will generally set a small fire which he feels he can control. This is because he is interested in having people who know him see him in action after the fire is discovered. Often he will discover the fire himself and call the fire department. Then he will warn or rescue any people who may be endangered by the fire. This gives him his chance to be a hero. When the fire engines arrive he will direct them in and try to assist the firemen in fighting the fire. Many times the fire will be small enough that he will have the fire extinguished by the time the fire engines arrive on the scene. This again makes him the hero. The firemen should be of great help to the investigator in the case of the attention-seeking fire-setter as they will usually have direct contact with him. They may find they have the same man trying to help them on more than one fire. This, of course, will bear investigation.

This type of fire-setter is looking for acclaim. He may be a quiet intraverted person who feels no one takes notice of him, therefore, he will set a fire so he can gain notice by doing a good job at the fire. The fire may be set near his girl friend's, or would-be girl friend's, home so she will see him in action and he can be her hero.

Sometimes a man may set a fire at some business where he wishes to gain employment as a night watchman. This will show the company the need for a night watchman and if he discovers the fire, it will show that he is on the ball and a good man for the job. Or if he already has the job of night watchman he may set a fire to keep his job if he feels the company is going to do away with the position. Or he may do so to gain a raise in pay.

There is also the case where a volunteer or paid fireman may set fires because he feels his company is not getting enough runs.

This man likes the people to see him riding the fire engine and fighting fires. This type of person will not usually set fire to buildings in which there are people.

THE "TORCH" OR PROFESSIONAL FIRE SETTER

This man is a professional criminal and an expert in his trade. He is for hire and takes great pride in his work. Most of the time he will work by himself, using timing devices and flammables. His fires are usually well planned, as he does not like failures. The professional may set more than one fire in a building.

The property owner who hires a professional will insure that he has an alibi for the time when the fire is set. He will also make it easy for the "torch" to enter the property that is to be burned.

In most cases, the professional will rely on "word-of-mouth" of those he has worked for to get him business. But, in some cases, they may even try to work with insurance men who could by their contact and who could suggest fire to their insurance clients.

Method of Operation

The "torch," like other criminals, can often be recognized by his method of operation. When he finds a successful way of doing his work, he will tend to stick to it. He will use the same starting devices, trails, and flammable liquids where possible. The professional fire-setter may or may not stay to see the fire. If he does stay, it may only be long enough to see that he has done his job well. But, in most cases, he will leave the scene as soon as possible and establish an alibi. He may leave town as soon as he has completed his job unless he has other work to do in the area.

Because the "torch" is a professional and it is part of his job to make the fire look *accidental,* he will go to great extents to do so. Therefore, the fire set by the professional will be one of the hardest to determine as arson.

SPITE OR GRUDGE FIRES

The spite or grudge fire is generally set at night. In this type of fire either a person's home or place of business may be burned. The fire will usually be set outside of the building, such as around the doors, window, heat or air vents, or under a porch. Often, objects of sentimental value will be burned. The spite or grudge type of fire setter usually likes to use flammable liquids in his work. The grudge fire setter may hold the grudge for many years before he sets a fire. The property owner, whom the fire was directed against, may be of little or no help when you question him as to who would have reason to set the fire, due to this great lapse in time. Possibly, the event which caused the hard feelings may have been so insignificant to him that he will not remember or will not feel it is important enough to mention. Therefore, it can sometimes be very hard to find the motive in a grudge fire.

Examples of Spite or Grudge Fires

The burning of churches and other religious buildings will often come under the title of spite or grudge fires. When there are religious or social problems in an area there will often be an increase in church fires. Often the person setting this type of fire will make no attempt to hide the fact that the fire was set. The person setting this type of fire is usually a fanatical person. In most cases, he will use a flammable liquid in his work; often they will use a home-made bomb or dynamite thrown from a passing auto.

If you have a series of fires which do not seem to fit into any other category, they may be the work of someone who has a grudge against the fire department. These fires will quite often be set at night or some other time that would be inconvenient for the fire department personnel.

We have probably all handled or at least read of a case where a woman has burned up the family car because she felt her husband ran around too much. This is a good example of the spite fire. Although we discuss the female fire setter separate in this book, it is plain to see that most of the fires set by a female could very well be listed under the title of spite or grudge fires.

In many cases, the fire set by the juvenile fire-setter will also prove to be a spite fire. It is possible that due to the fact that his parents have denied him the use of matches and perhaps punished him for having same, he will take matches when they are not looking and set a fire. When questioned the parents will state that their child could not have set the fire, as he is not allowed to have matches.

A man who has had a local store in the community for many years may find that a new store of the same type has moved in down the street. This new store, by having a modern building and giving better service, may soon take his customers and place him in financial trouble. In this case, he might decide that the best way to handle the problem is to burn out his competitor. This is not an uncommon problem for fire investigators, and we must also realize that it can work in reverse.

In dealing with spite and grudge fires, as concerned with businesses, one of the most common problems is the employee who has been fired, or the subject who was turned down when he applied for a job. These two types of people are often the cause of spite and grudge fires in business places. Also, we must not forget that a disgruntled employee may set fire to the place where he works.

INSURANCE FIRES

The man who sets the insurance fire will, in most cases, be a respected member of the community. He will have no past criminal record. Here we are dealing with the individual or businessman who sees need for a fire, but does not have the money or contacts to hire a "torch" to do his work. The insurance fire setter, in most cases, uses flammable liquids and a timing device. This man wants an alibi for the time of the fire.

Reasons for Insurance Fires

Often the auto fire will be an insurance fire. The owner may find he is faced with expensive auto-repair bills and he will then decide it would be cheaper to have his auto burned and he could collect his fire insurance. His downfall, at this point, may be the fact that he will remove the accessories from the car and all valuables from the trunk and glove box. Sometimes, the fact that a car is about to be repossessed can make a man decide to burn it.

If a man is forced to move to another location or state, perhaps due to a transfer of work, he will find he must sell his home quickly in order to have money to relocate. If he cannot get a quick sale or the proper price, he might feel that he is better off to have a fire.

There is also the case of the man whose property was condemned by the local government due to a new highway that was to pass through this property. He decided the government would not pay him a good price for his property so he burned his home down to collect the insurance.

The businessman can find many reasons to have an insurance fire. He may find it is necessary for him to remodel his place of business to meet competition and having a fire may be the cheapest way of doing this. Often, when having a fire, the businessman will remove good stock and replace it with cheap stock. After the fire, he will make his insurance claim for the good stock and then sell it on the side. This is one reason an examination of the ashes of the stock should be made.

Sometimes a businessman will find he is loaded with outdated or outstyled stock. This stock can only be sold at greatly reduced prices, at which he will lose large amounts of money. Here again he may turn to fire as a way out.

In many cases, the fire setter will remove any personal items of value from the building and any items of sentimental value. They also will often remove records from their normal place in order to protect them from the fire. This is especially true of fire insurance papers.

CRIME COVER-UP FIRES

This person sets fires to cover up other crimes he has committed. Here the type of crime the person is trying to hide will tell you more about the type of person. The criminal will generally start the fire in the part of the building that was the center of the first crime. This person is not a professional fire

FIGURE 27. Here we have a picture taken in a home hit by an arsonist-burglar. You are looking into a bedroom clothes closet. Clothes and other items piled on the floor in the closet were lit and the fire then spread to the clothes that were hanging. There are holes burned in the floor and the rear wall of the closet at the floor. You can see by the markings on the rear wall how the heat and flames traveled up from this point. By looking through the hole burned in the floor you can see that the floor joists at this point are also charred. There are some pieces of plaster board lying on the floor in the closet. Some of this was knocked loose by the firefighters while they were doing their job. As you can see this type of fire caused a great amount of smoke and soot. There is little doubt but that this fire was set just to cause the smoke and soot in the hopes that it would destroy any fingerprints or other evidence that was left on the scene. We might add that there was very little actual fire damage but there was a great amount of smoke damage.

setter so, in most cases, it will not be too hard to tell the fire was set. Usually, due to the improper setting of the fire, all evidence of the first crime will not be destroyed.

If the first crime was embezzlement or some other financial irregularity, then the fire will be started near the books. In most cases, the books will be open and may be standing on end instead of lying down. A book may be opened to the page that shows the error. In any case, it is found that books do not burn very well. The type of first crime here would tell us that our fire setter is a white-collar type.

A murderer will set a fire hoping to completely destroy the body so no trace of it will be found. Or he may hope to mar it enough to prevent identification. In some cases, a fire may be used hoping to make the death appear accidental. Very often a flammable liquid will be used in this type of fire. The origin of the fire will be near the body. As the subject has already committed the most serious of felonies, he will not hesitate to commit arson.

The person who will commit burglary or breaking and entering usually does not commit arson because he realizes it is a serious crime and does not wish to take the chance of being caught and tried for both crimes. But, of course, there are those who will take the chance. In these cases, they set the fire in an attempt to destroy any evidence they may have left at the scene. You will find that because of the fire, smoke and water, much of the evidence to the first crime may very well be destroyed.

SUMMARY

1. Juveniles will most often set fire to fields, woods or vacant buildings. In most cases, these fires will show a lack of expertness in the way they were set and the fires themselves will not make sense.

2. In most cases, the juvenile will use matches to light paper or any other burnables that are on the scene.

3. An attempt should be made to question all juveniles who are at the scene of a fire.

4. A juvenile may set an "expert fire" and even use a timing device, even though, as a rule, they will not.

5. The area should be checked for physical evidence, such as matches or toys left on the scene.
6. The female fire-setter is usually an upset woman; in most cases, she will be mad at a man.
7. The attention-seeking fire-setter will set his fire in an area where he is known because he wants to be seen fighting the fire or rescuing people; he may be a shy, backward person who feels that no one notices him.
8. The "torch" is for hire and he is an expert in his trade. He will use a timing device and flammables and will often set more than one fire in the building.
9. The "torch" will go to great extremes to make the fire he sets look *accidental*.
10. The spite or grudge fire will often be set outside of the business or home of the person toward whom it is directed. The hours of darkness are a favorite time for the setting of this type of fire.
11. Flammable liquids are often used by the grudge or spite fire setter.
12. Spite or grudge fires can be directed at individuals, businesses, churches, fire department or any other group.
13. In most insurance fires flammable liquids and timing devices will be used, as the fire setter will want to do a complete job and still have time to get an alibi. In most cases, he will be a respected member of the community. Often important papers or objects of sentimental value will be removed before this subject sets his fire.
14. When a fire is set to cover up another crime, it will be set in the area where the first crime was committed, in an attempt to destroy any evidence. This person is not a professional fire setter and, in many cases, will not do a good job.

Chapter Seven

TRACKING DOWN THE ARSONIST

AFTER the investigator has determined that a fire has been deliberately set he should have some idea from his investigation what kind of fire setter he is seeking. He may suspect it was set by a professional, by the honest citizen for insurance, or by some kind of nut for a personal satisfaction.

PYROMANIACS

We stress this type of arsonist because of his danger to human life. The pyromaniac, who suffers from a progressive illness, is especially dangerous in his advanced stages because he has no regard for human life.

Lack of Motive

The lack of motive is a trademark of the pyromaniac and the investigator can use this knowledge to his advantage by directing his suspicions toward this type person when investigating senseless but apparently deliberately set fires. If there are a series of these fires you could compare notes from reports of other fires and try to match descriptions of persons or vehicles seen in the area. This again is where intelligent questioning of witnesses is important. Examine photographs taken at the scene which include faces of on-lookers and see if anyone appears to be getting gratification from the fire. A plain-clothesman mingling with the crowd can also be helpful at this type of fire.

Consult the records and the memories of fellow officers especially if you have some kind of description of a suspect or vehicle. The suspect may not have a record for arson or he may not even have a record at all, but he probably was questioned for something which does not seem to be related to arson. Perhaps he was questioned about being a prowler or "peeping Tom" or even about larceny from a clothesline, where the complainant would not prosecute. In these latter cases a good memory of the officers on the post can be important.

The investigator should examine the scene of the fire carefully for an article of clothing or other clues left by the pyromaniac because, while he may not have fear of being caught, he is internally excited at the thought of satisfying himself and may stumble or even be in a minor accident with his vehicle while leaving the scene. While footprints and fingerprints are important in crime detection, they are not often available at a fire scene where some thirty firemen trampled the area, and smoke and heat obliterated all clues. On occasion, in an area plagued by arson, the authors have made special effort to arrive at the fire before the firemen to protect any clues, such as footprints, which tell a great deal about the suspect. A footprint can be protected temporarily by placing a trash can or some lumber over it. Then, as soon as possible a cast of it should be made before it is washed away by the water from the fire hoses. While fire department cooperation is excellent, the authors do not recommend asking firemen at the scene of a fire to be careful not to step on any footprints.

To further narrow the field of suspects, many persons in the area, including both those on foot and riding, should be questioned and a record kept of their statements and their names and address. This can be compared with reports of investigations of other fires. If the fire was set by a pyromaniac you know he was in the area and most likely still is or is at some vantage point where he can get the full enjoyment of the spectacle he created. Enlist the aid of other officers in the area to get the necessary information from possible suspects they may observe.

If the investigation leads to a good suspect soon after the fire, the line of questioning toward the pyromaniac should be that of his needing help and the investigator being in a position to help him. After the pyromaniac has had his satisfaction from the fire, his mental state may change to where he realizes he needs help and this will put him in a vulnerable position for

FIGURE 28. This, as you can see, is a modern one-story brick home. This home had a very hot fire that was confined to one room and which burned through the ceiling and into the small attic area above it. It then burned through the roof. The one firefighter is standing in the small attic area with his body up through the roof. The fire was confined to the one area of the house and the rest of the house had only heat and smoke damage. This fire was found to be of an incendiary nature. The subject was arrested but not before he set other fires.

the alert investigator. Any previous offenses or evidence against
the suspect can be brought out at this time. Notify the suspect
of his rights and have the statement blanks ready now, as tomorrow he may recompose himself, or his lawyer may help to
recompose him. This subject is a menace to society and the
message should be gotten across to him that unless he is helped
immediately, the next time he will be in much more serious
trouble. The investigator can strengthen his case now by getting
information or material from the suspect as to how and where
he started the fire. Information that only the person who started
the fire would have is good evidence. If the arsonist used an
accelerant to start or spread the fire, residue will invariably be
on his clothing or hands. If he used book matches and cast them
aside on the ground while starting the fire these should be gathered up and can be microscopically examined and associated with
a partly used book found on the suspect. It may take a number
of fires set by a pyromaniac before the investigator can come
up with a suspect. Keep well informed of all the statistics of
these fires and try to develop a similarity in the pattern. Possibly
a time factor or increase in severity of the fires in the case of
pyromaniacs. In series fires, a good investigator can almost tell
when and where the next fire will be, and should prepare himself accordingly. It may pay off to stake out an area and have
the officers keep a record of suspicious persons and vehicles but
not to contact anyone. This is not because you will scare the
pyromaniac away and prevent a crime; you will just chase him
to another area. This, of course, is not necessarily true of other
types of arsonists. Remember, you are dealing with a mental case
who may have at one time been institutionalized and his thinking
will not be the same as that of the cunning criminal. He may
turn up anywhere, and may even stand there watching you make
your investigation, or may approach you and ask questions about
the fire. As a matter of fact he could be an "average citizen"
most of the time and periodically become possessed with his desire to set fires. However, the case history of this type of person
always reveals some type of maladjustment. Often the guardians
or relatives of a pyromaniac will move him to another area,
thinking a change of environment will be helpful to him, es-

pecially if he is in frequent trouble with the police in his home town. This makes a difficult situation for the fire investigator and any suspects or persons questioned who claim to be newcomers to your area should be recorded and a criminal arrest record obtained from the authorities at the previous residence of the suspect. Also notice if the date of arrival of the suspect in your area coincides with the first of the particular type fires you are investigating. In most cases where the guardians or relatives of a pyromaniac have moved him from area to area they will, when first questioned, deny that he is a fire setter. But, in most cases, when confronted with the facts and an understanding that you wish to help him, they will admit his problems to you.

We must remember that the pyromaniac is mentally sick and cannot help himself and we as investigators must treat him as such when we make our arrest.

Motives

In cases of arson, or deliberate burning, we must remember that monetary gain does not have to be the motive for the fire. Since establishing the motive for a suspicious fire can be the most important phase of the investigation we must familiarize ourselves with all possible motives. If the management was involved in labor troubles, his home, business, or car may have been burned by a fanatic from the opposition. This can include bombings and bomb threates as fanatics often resort to these extreme measures to stress their point. Religious, racial or political disputes can be involved in a deliberate burning and an investigator should be informed as to the existence of any of these conditions. It can be mentioned here that in the aforementioned cases a deliberate burning is never an organized plan of the victim's opponents, but rather the work of one, or a few members, who decide to take matters into their own hands. Business competitors as well as customers who feel they were cheated by a business man can have what seems to them a valid reason to "burn a man out of business." This can include car dealers, television, auto or appliance repair shops, restaurants, etc. If a man goes into business in an already crowded area and the customers flock to him because of a better product or price, the livelihood of his competitors may be threatened. These facts should be taken into consideration when his new shop suddenly burns to the ground.

An investigator can explore the condition of a victim's home life. Domestic problems often enter into deliberate burning for spite, hate or money. Persons who are separated, divorced or just living with someone else are especially vulnerable to this type of burning. One person could be jealous of the other who has taken up a new life and made a social spectacle of the other. A man may be behind in his alimony payments; or perhaps he purchased a new car to take out his new girl friend, which may enrage his ex-wife. Her vengence will be satisfied when she sets fire to the car some night while he sleeps and the new car is parked in the drive.

Rejected suitors, as well as married, divorced or separated persons may feel they have reason to set a fire. This motive, however, will most likely be jealousy. The victim of this type of fire can tell the investigator much about the mental stability of his or her lover, and if there were any previous displays of vengence.

An investigator tracking down an arsonist, even trying to determine a motive, may feel sometimes that he is groping at straws. Clues at the scene and ready information from witnesses will be difficult to obtain. You will have to know how to make the most out of the slightest clue. A used matchbook or match, a thread of hair or material around a broken window, a footprint or an article of clothing left at the scene by a careless fire-setter can give all kinds of information when properly analyzed. Clues on a flashlight left at a fire scene may have been obliterated, but the smooth batteries protected inside can produce excellent fingerprints. A complete examination should be made on all articles found at a fire scene for names or initials of the owner and the trade names of the manufacturers. Be particularly alert for hairs, laundry marks and papers or belongings in the pockets of any clothing.

HIRED FIRE-SETTERS

The professional fire setter will require the same procedure as any other criminal specialist. Like any other well-known criminal, he will have an extensive record and will be well known to the underworld. He may have had his latest successes in another state so it is well to check with other authorities for similar fires when suspecting a professionally set fire. Tracking down the professional can have the advantage of having more than one person to apprehend: the persons who paid the "torch" and will benefit from the fire are equally as guilty and must be charged.

The Beneficiary

It may be easier to work on this angle first, as this person will not have the experience of the "torch" and is more likely to make a mistake or fall for a trap set by a clever investigator. Check this man's financial status and determine if he made any large withdrawals lately from his bank account. The "torch" will demand a large amount of money before the fire, and will often settle for the rest and a bonus after the fire. He may then leave town, but the suspect who benefits from the fire will have to remain in order not to arouse suspicion and in order to collect his insurance—if that was the motive for the fire. Collect as much information about this subject as possible, such as the condition of his business and any outstanding debts he had. Talk to other business persons and determine if the suspect made any statements or remarks that even referred to his being put out of business or to some type of misfortune. Advertising for a fire sale may have been prepared before the fire. Any rough sketches or information indicating a going-out-of-business sale can be collected for evidence. If inventory sheets are available check them and determine if more or less merchandise than normal was on hand; also the quality of merchandise on hand as compared to normal. A lot can be learned from these records and comparing with the insurance claim. In a "torch" set fire the one who will benefit will be greedy and want all he thinks he can collect to offset the cost of having someone set the fire. Being a businessman he will figure in dollars and cents and decide he has to make up the loss by padding his claim. He may even claim some merchandise was destroyed that he had hid and intends to sell later to some other crook. Merchandise inferior to that mentioned in his claim may be burned. The investigator can check this by careful inspection of the ruins. Most quality merchandise contains some parts or different material than its cheaper counterpart. This is true of furniture, shoes, clothing or any other merchandise. The clue may be screws, nails, brackets or chemical analysis of the ashes that would indicate the presence or lack of substance in the burned material. Close cooperation with the manufacturer is needed and he may supply you with some parts or accessories from his product which do not readily burn and should have been found in the ashes. He could also tell you if parts of merchandise you found came from his factory. While you are talking to this manufacturer you can determine from him or his delivery man if the suspect had any merchandise delivered in any manner other than normal or was advised to leave it on the sidewalk or asked not to bother carrying it to where he usually does. All these pieces of information when put together will help you build your case. Be congenial to the manufacturer you have contacted as he is expert in his trade and knows as much about his competitors' products as his own and can be a big help to you in identifying clues you found in a fire.

Employees and Bookkeepers

Most firms have a full- or part-time bookkeeper who should be questioned. He may say the books were burned and he can give no information. People who work with figures have good memories and he will have some idea of the recent transactions of the firm and its financial condition. Have him list as much as he can remember. He may have to contact your suspect to find out how much he should remember. If the bookkeeper is not part of the crime he can be one of your biggest assets if you gain his confidence. Other employees of the burned-out firm must be completely interrogated as even if they were not part of the plot they would have noticed any change in the attitude of their employer, the normal routine, or mysterious visitors. The employee will have firsthand information about the amount and quality of stock on hand as compared with normal. He can also advise you if the stock was stored in any manner other than usual. If any fire doors were not in working order or blocked by stock. It is possible that one of the employees was charged with seeing that the building was locked at closing time, but on the date of the fire the boss told him to go home and he would lock up. A check should be made to see if the company has or had a night watchman. He may have been given a few days vacation, during which time the fire occurred or he may have been fired for some odd reason shortly before the fire. Here we are getting back to the point that generally when a professional fire-setter is hired, entry to the building is made easy for him. Check with people living or working around the building and see if the owner or anyone else has been seen entering or leaving the building odd hours of the night or on weekends.

Assistance From Other Officials

Speak with the policeman working the area as he will usually know who, if anyone, is usually in the building at odd hours. He may also have noticed any odd shipments going or coming from the place. A check of police records will also show how often, if ever, the owner has forgotten to lock the doors of his place of business. Some places often have doors left unlocked and others never have a door left unlocked. We must remember that even if a door is left unlocked for the "torch," he will, in some cases, do some damage to make it appear that the place was broken into. This is done so if the fire is a failure, it will appear that someone broke in to steal something and then set fire to destroy the evidence. But some professionals are so sure of themselves that they will not show doubt in their work by doing this. In most areas, the police are required to make checks of the business places during the night, so check with the officer working the area and find out what time he made his last check. If everything was in order at this time, compare the time the fire was reported with this time. If the fire started after the officer made his check, could it, under normal conditions, have made the headway that it did in this period of time? A check of fire inspection reports made by the fire department in the past can be helpful in making this decision. We will also want to know if the fire started in one of the danger areas noted in the inspection report, or if it started in the area that was not noted as a danger point, but that was hidden from sight of the officer making his rounds or anyone passing by. Talk with the fire department officer who made the inspection report; he may be able to give you some information that is not in his report. Find out how he was treated when he went to make his inspection, was the owner disturbed at his being there or did he treat him overly friendly. Did he find any fire safety violations and were they corrected? Did the owner correct them faster than usual? Or did he offer the fire officer a bribe to forget about them? Any way that he was treated different than normal will be of interest to the investigator.

Search For Witnesses

As in all other cases, we will want to question anyone who was in the area of the offense and may have seen or heard something. If the fire occurred during the day, in most cases we will have many people we can question. When the fire occurs during the night, we will have fewer people to question, even if they can give us no other information, they can help us pinpoint the time of the fire, by telling us the different times they passed the building. Most business places will close early in the night, but the bars in many areas stay open until 2:00 A.M., so we can question any bar patrons who pass the location of the fire on their way home. In some areas there will be taxi cabs working during the whole night. A check of their records will show if they had any fares in this area, and a check with the drivers may show that one of the drivers, while just cruising around, passed through the area. Milkmen and papermen may start work anywhere from 3:00 A.M., on, and a check with them can often be very helpful. Depending on the area, there may be other people working, or going or coming from work, during the night. Every area has some local "bums" who are out all hours of the night, but who would not commit this type of offense. Question these people, they may have seen something and in many cases, they may be more than glad to help you.

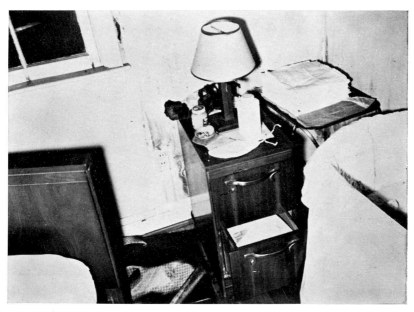

FIGURE 29. This picture shows an open drawer in a piece of furniture in the bedroom. This was just one of the many signs that this home had been ransacked. The dark streaks on the walls will give you an idea of the smoke and soot damage done to this home. As you can see the window panes are also heavily streaked. The burned cloth laying on the floor between the night table and the bed is the curtain from the window shown in this picture. Investigation revealed that this curtain was pulled from the window by the firemen that put the flames out. This same fireman moved the bed to get to the curtain. By arriving on the scene while the firemen were still there it was easy to gain this information. If we had arrived later it would have been very difficult to determine if the curtain and the bed had been moved for some reason by the arsonist, or if they had been moved by the firemen. This picture also shows that the investigators were careless with their used flash bulbs. There are two in the picture, did you notice them?

Insurance Records

Find out what company insures the building. Contact the insurance company and they will be able to advise you as to just how much insurance is on the building and contents. Also, they can tell you if there have been any recent changes in the coverage on the building or contents. Any increase in coverage would bear looking into. If the insurance company has an investigator of their own working on the case, you may want to compare notes with him, as he may have some information that you do not. He is trained in insurance work and will be familiar with the different ways of trying to beat the insurance company. You may even find that the owner has several policies with several different insurance companies on the same property. In this case, one insurance company may not know of the policies held by other companies. A check of insurance company records may show that the subject has had other fires in the past. These fires may have been at another location outside your jurisdiction, so your own records would show no record of them. In this case, you would want to know if in these other fires there was any question as to whether the fires were *accidental or incendiary.* Gather any information you can on investigations made of these fires. The insurance companies should be more than glad to give you any of this information they have and you can contact the local authorities in the areas where the other fires took place, for any information they may have. They may be interested in the information that you have gathered in your present investigation. Compare all of the fires for any similarities, because, a "torch" who has developed a successful method of doing this work will not usually change his method. Also, if a man has had success in dealing with one "torch" in the past, he will, in most cases, hire the same man for any future work he has. Similarities of the fires will therefore suggest they were of incendiary nature and that they were set by the same person; if not by the owner then by the same "torch."

Bank Check On Suspect

Keep a close watch on the owner after the fire, since in most cases he will still owe the "torch" his final payment. For this payment personal contact is not generally made. If the owner did not withdraw the money to cover this at the time of his first payment, then this will show as another withdrawal from the bank. A check should also be kept on his bank for any large deposits. Since his business has burned (if this is his only business) he should have no income unless he has sold some items on the side that were supposed to have burned in the fire. If he rebuilds his business within a short time, check and see if he has done so with less debt than would be normal under his supposed financial condition. This is often a sign that the owner has made money on the fire.

SYNDICATES AND ARSON RINGS

Convicting an arsonist who is a part of a syndicate or an organized crime group will require a tight case. This the authors have found to be true in all types of crimes where the suspects were believed to be part of an organization. The financial backing, expensive lawyers and the ability to produce alibis and witnesses by a suspect are all signs of an organization. A fire set by a syndicate-backed arsonist will be complete and usually will have the insurance as the motive. By the time the fire is out and the investigation begins, the "torch" can be well on his way to another state to do another job. This poses a serious legal problem for the local investigator. Attorneys for the Justice Department are aware of this situation and Congress will be asked to amend the laws so the FBI will be able to intercept fire setters in interstate travel with the purpose of furthering their unlawful activities. Arson for insurance is a natural for the syndicate whose other rackets may be faltering due to federal or local exposition and pressure. Businesses owned secretly by the criminal element that are not among their best money-makers will be among the first to go up in flames, when some quick ready cash is wanted. An investigator must make every effort to determine the real owner of a business, and obtain all possible insurance records when a syndicate is in evidence during an arson investigation. A criminal record check of all persons connected with the burned-out business will often give a clue that organized crime is behind a fire. Since organized crime operates on a national scale, the actual "torch" in an arson fire that is syndicate backed may be the only one you can build a case against. A copy of your report, however, should be sent to federal authorities or any other agency it may concern.

MAKING THE ARREST

Let us remember that when we are dealing with the "torch," we are dealing with a professional criminal, a man who knows the law and knows just how little he is required to tell you. So be sure you have a good case on him before you pick him up. Do not expect to pick him up like you would some kid who has set his first fire, and shake him up and have him admit everything to you. This man has been around, in most cases; he has been arrested before and knows the ropes. Since this man is usually proud of his work, it will sometimes help him when questioning him to start downing the work of the fire setter. You may even want to talk about what a sloppy job was done in starting the fire and how easy it was for you to find evidence and locate the cause of the fire. This will hurt his pride and sometimes cause him to say something that will be of use to you. No one can tell you just what to say when questioning a person or what line of questioning will give the best results. The line of questioning that will work on one person may not work on the next, and what will work for one investigator may not work for another. You will find through trial and error which is the best method of questioning for you to use on which type of person.

SUMMARY

1. Determining the motives or the lack of a motive in an arson fire will be your best clue in helping to identify the type of arsonist involved.

2. Be alert to company books and records left in a position to be consumed in the fire. This could indicate embezzlement.

3. . Criminal record checks on all employees and officers of a burned-out business should be regular procedure when arson is suspected.

4. Learn how to question suspects. A composed suspect will give very little information. Find a topic that will put him ill at ease and never tell him you have no clues.

5. The odor of smoke or traces of an accelerant on a suspect brought to his attention by the investigator can do much to upset the composure of a fire setter.

6. If you suspect a professional fire-setter is involved, it will be better to build the case against him by breaking down the less experienced principal involved. The "torch" will be in hiding or will have left town and will not volunteer himself for questioning, so you will have to wait until you have enough evidence to procure a warrant for him.

7. Do not make a premature arrest. Be sure all evidence and statements were properly handled. Have an assistant with you at all times when questioning suspects to cosign all written statements and bear witness to all oral statements. The authors have seen oral statements carry much weight in court, when made by a suspect to a properly prepared investigator.

Chapter Eight

VACANT BUILDING FIRES
HOUSES

FIRST we will consider vacant houses. As always, it is important that the investigator arrive on the scene as soon as possible. On his arrival he should note the time and the extent of the fire. If the fire is still in progress he should take photographs of the overall fire-scene right away. Any odd odors or colors of smoke should be noted and identified if possible. The fire department officer in charge should be questioned as to whether he has noticed anything odd about the fire. The first fire department officer and men to arrive on the scene should be questioned as to the time they received the call and the time they arrived on the scene, the extent of the fire on their arrival and any other information they can give you about the fire. Also question them as to who was at the scene on their arrival and whether they saw anyone leaving the area as they arrived. As all fires cause some sort of crowd to gather, walk among the crowd and listen to the people talk, question anyone who seems to know anything about the fire. Be sure to get names and addresses. Watch for known fire setters or anyone who seems to be enjoying the fire. Contact the fire department headquarters and attempt to get the name and address of the person who reported the fire. Question this person as to the time they first noticed the fire, the extent of the fire at that time, anyone they saw in the area, and any other information that they may be able to give you.

Protect Evidence

You may not be able to enter the building when you first arrive, and you may become involved in the above procedure. It is therefore important that you enlist the aid of the fire chief;

FIGURE 30. Houses such as the one shown here are often a problem to the fire departments in rural areas. Because of their age and construction, they will burn very readily. Many times they are located in lonely areas with no other houses nearby; they may be vacant and wide open for children or any others who wish to enter and set fires. Due to the above conditions, the fire will have a good start before the fire department is notified. By the time the firefighters arrive on the scene, they will find conditions at least as bad as you see here. These advanced conditions and often a lack of water supply at the scene will give the investigator a pile of ashes to investigate.

advise him to notify you as soon as it is possible for you to enter the house. Also ask him to advise his men to touch nothing more than is necessary for them to do their work. If they must move anything, if possible, ask that they let you photograph it first. Also ask that no clean-up work start until you have finished your investigation. Remember that no evidence should be moved until it has been measured, photographed and identified.

Investigation

In most vacant houses the gas, electric, and heat will be turned off, check to see if this is true in this case. If everything is turned off then there is very little chance that the fire started without human help. Lightning would be the next most likely cause. Of course, if there has been no electrical storm we can rule this out.

We will follow our usual procedure and locate the point of origin of the fire. Now we must locate the cause. A close search of the point of origin must be made. We will look for timing devices, matches, or signs that flammable liquids were used. Very often we will find that there are signs that someone has been living in the house. In this case, it may be that a vagrant or bum has made his home there, that young children are using it as a club house or that teenagers are holding parties there. If clothing is found, this will help tell you what sex and age bracket has been using the house. Children will usually leave candy wrappers, soda bottles and toys laying around. Beer bottles, playing cards and used contraceptives would be a good sign that teenagers have been using the house. Wine bottles, old clothes and a razor would point to a vagrant or bum. Often the police who work the area will be able to help as they may have in the past chased children or teenagers out of the house. A check with the people who live in nearby houses will often give information as to who has been seen coming and going from the house. In the case of a vacant house, it is very important to speak with the neighborhood children as, in most cases, they will know better than anyone else who is using it. If a bum is living there they will know it and if teenagers or children are using it they can often give you names. Most pre-teenagers will help you if

FIGURE 31. This picture shows the point where the fire started in this building. The floor boards are completely burned away in the area nearest as you look at the picture. The walls were of plaster over wooden laths construction. The fire started in the area to the right of the window. The wall area to the right of the window not only has the plaster missing but the laths are also burned away almost completely. The wall behind the fireman has almost all of the laths still intact. The window sill burned fairly well due to the oxygen received from the window. If you look close you can see some of the alligatoring in the window sill, floor joists and the wall studs. The window frame is missing from the window. This was removed by the firemen in their clean up work.

they can. If anyone is using the house then it is possible that the fire was an accident. When there are signs that someone may have been living in the vacant house, it is very important that a good search of the ruins be made as the subject may have perished in the fire. As you gain experience in the fire investigation field you will learn to know the smell of human flesh burning. Fire fighters sometimes smell this odor while fighting the fire, and may come to you and advise that they believe someone burned in the fire. If you have reason to believe that someone was in the fire, notify the fire chief and he will have his men assist you in making a search of the ruins.

FIGURE 32. Here we have a case where the fire setter broke a hole through the inside wall of the house and dumped a flammable liquid in between the walls. You can see the can that held the flammable liquid is on the floor in the picture. As you can see, the fire burned right through the outer wall of the building. It was only through quick work by the fire department that this building was saved. The building was very old, of frame construction, and its general condition was very poor. The case was cleared within a few days.

Type Of Fire Setters

In many cases, vacant-house fires will be set by juvenile fire-setters. This would be especially true if the house set alone in a wooded area. (The authors have also had cases were young juveniles set fire to vacant houses in a housing development in broad daylight.)

FIGURE 33. This photo shows the burned-out corner of a two-story frame house. The fire started in the corner of a room on the second floor. The house was vacant and there was no furniture in it. The gas and electricity were turned off. A check of the window in the lower front of the picture will show that many of the window panes were broken out. This was not caused by the fire. It was done by children playing in the area. The fire traveled up through the walls and into the attic. The small window in the peak is in the attic. You can see that much smoke came out of this window. The eaves near and at the peak show a great amount of charring and burning. This fire was started by juvenile fire-setters. The fire was set in the middle of the day. There was no school this day.

We must also consider the fact that the vacant-house fire can be the work of a pyromaniac. Very often, especially in their early stages, they will set fire to vacant buildings. They can set fire

FIGURE 34. This home was being rebuilt when it was destroyed by fire. By examining the front of the building you can see that it was of clapboard construction. In the right front of the picture there is a pile of building supplies for this building. Because the building was open, ventilation was no problem for this fire and it burned very fast. On the front of the building you can see a good example of the checkering of the boards by the fire as it is burned.

to the building with little chance of being seen and yet it will make a nice fire.

A check should be made to see if the vacant house is insured, as the fire could be an insurance fire. It is possible that the house is vacant due to the need of expensive repairs, and therefore a fire may be the best way out for the owner. Or it is possible that the owner had to move to another area and is having trouble selling the house for the price he wants. Therefore, a fire will get him the money he needs. He may even try to claim insurance on furnishings that were not even in the house. In some cases an owner may burn down a vacant house rather than to go to the expense of tearing it down.

BARNS

In most cases of vacant-barn fires we will find that there was little or no hay left in the barn when it was abandoned. This plus the fact the barn may never have had a heating or electrical system leaves little reason for a fire to be started in any way but by human hands whether it be *accidental* or *incendiary*. Very often, oil-type lanterns will be found around a vacant barn and, in many cases, there will still be some oil in them. Children playing in the area may find them and, of course, try to light them. Sometimes they will use any hay that is left to build a campfire or they may try to smoke the hay. Of course, hay and oil lanterns are a great aid to any would-be arsonist in the area. Due to the fact that many vacant barns are located in areas that are not heavily populated, they will often be burning intensively before anyone will notice them. Also, it is not very likely that anyone was seen coming to or going from the barn and this will make the investigator's job much more difficult. Our system of investigation will be basically the same as that used in the vacant-house fire.

FACTORIES

When we are dealing with vacant factories we find that we have some special problems that we did not have in the other vacant buildings. Often the so-called vacant factory will still have valuable machinery in it. Naturally machinery means oil and grease which will be an aid to any fire that starts. Due to the nature of the machinery it may be necessary that there be heat in the building and electricity. In many cases, electricity may be left on in order to keep a burglar-alarm system working. Even if there is no machinery left, there will be the poor house-keeping problem. You will usually find that piles of oily rags, papers and cardboard boxes have been left behind. Often there are broken windows, and windows and doors left unlocked, which make it easy for anyone to enter the building. Due to the expense of a factory building, there is a good chance of an insurance fire. This building is, of course, a sitting duck for the juvenile fire-setter or the pyromaniac. The vacant factory will, of course, be a drawing point for children who are looking for an interesting place to play. Children of various ages will find a vacant building of any type to be a good place for their club to meet. Teenagers will use this type of building for beer and sex parties. They will smoke and be careless with their cig-arettes and matches. If the weather is cold and there is no heat in the building, they may even improvise their own heating system.

Juvenile gangs will often use this type of vacant building to store stolen items until such time as they get a sale for them. Often they will accumulate quite a pile of stolen goods, much of which may be burnable. Now this once-vacant building has goods stored in it, and people coming and going at odd hours who may accidentally start a fire. The possibility of a fire in this vacant building has certainly grown.

Naturally this type of building is also ideal for the bum, hobo, or drunk who is looking for a place to sleep. In cold weather this type of person will very often build a fire to keep warm. This fire, along with the poor housekeeping in the building and perhaps the drunken condition of the "new tenant," will make conditions perfect for a fire and possibly the loss of life.

SUMMARY

1. Question the fire department officer in charge as to any oddities noticed in fighting the fire, such as, odd smells or odd-colored smokes.
2. Enlist the fire officers in protecting evidence at the scene.
3. If the gas, electricity and heat are turned off in a vacant house, then there is little chance of a fire starting without human help.
4. Locate the point or origin of the fire and look for physical evidence.
5. Look for signs of someone living in or playing in the building. The neighborhood children will often know who uses a vacant building.
6. If there is evidence that someone was living in the building, then a search of the ruins for a body is in order.
7. The vacant-building fire may be the work of a juvenile fire-setter or a pyromaniac.
8. If building is insured, then there is the possibility the fire was an insurance fire.
9. Vacant barns seldom have gas, electricity, heat, or enough hay in them for a fire to start without human help.
10. Electricity may be left on in vacant factories.
11. Machinery, oil, grease, oily rags and poor housekeeping will cause special problems in vacant factories.
12. In the case of a fire in a vacant factory, insurance should always be considered as a possible motive.

Chapter Nine

MOTOR VEHICLES FIRES

FIRES in cars and trucks are common and it is almost always necessary for the investigator to make a complete report, at least for insurance purposes. Unless the vehicle burned completely, as in the case of an accident, it is not too difficult to determine the general area where the fire began. The firewall divides the engine compartment from the passengers and usually contains the fire until it is extinguished. A fire in the engine compartment can often be traced to a leaking gasoline system and oil-soaked wires are frequent causes of fire under the hood. A fire in the passenger compartment normally would indicate an upholstery fire, probably from a cigarette or spark. Shorted wires under the dash are not a common cause of upholstery fires in cars as they burn themselves out before they can do any further damage. Motors being cleaned with flammable liquids often are ignited when the metal part of the cleaning brush causes a spark. Notice if the battery was disconnected and removed before the cleaning was begun. Cargo carriers often carry flammables that have to be properly stowed and grounded. A motor vehicle fire could have been started accidentally due to a leaking gas tank or fuel pump. Perhaps a passerby casually tosses a cigarette in the gutter or the owner may start the vehicle causing a spark from the ignition or the exhaust to ignite the fumes from the fuel on the ground. The investigator would, in this case, question the owner of the vehicle to determine the condition of the vehicle before the fire. Examining the gas tank may reveal that it has been damaged or dented from hitting a curb or other object which may have opened the seam. Of course, if the gas tank exploded because of the fire there will be dents where the tank struck other parts of the car upon exploding. Notice the po-

FIGURE 35. The station wagon shown in this picture was involved in an accident with another vehicle. At the time of the accident the gas tank ruptured and the station wagon was engulfed in flames. Inside the front door of the station wagon you can see the leg of one of the occupants of the wagon. They had no chance to escape. You can see the damage to the left rear of the vehicle where it was struck. You might note that the left rear tire is completely burned away. By looking in through the windows you can see the wire springs and metal framing where the seats are burned away. The paint on the outside of the vehicle is completely discolored.

sition of the car and question the owner as to when he filled the tank. He may have damaged the tank a week or a month ago and did not know the tank leaked because he had not filled the tank to the seam since it was damaged. A leaking fuel pump on a parked car can usually be traced to a defective bowl gasket or flex-line on the pump. A vehicle with either of these fuel-pump defects will leak gas for a considerable time after being parked depending on how the position the car is parked affects the gravity ratio between the fuel pump and the gas tank. The position of the car can be taken into consideration when investigating this type of fire. The investigator can note the contour of the road and in which direction gas leaking from the fuel pump would flow. If, as an example, the road was level except for the contour to the curb, and the fuel pump was on the curb-side, you may find the tire and brake hose burned as well as engine parts around the fuel pump such as the radiator hose and wires. Of course, the sooner the fire is extinguished the easier it will be to determine the cause.

Trunk Fires

Many people carry flammables in the trunks of their cars which can be dangerous even when in air-tight cans. It is not unusual for a can carried in the trunk of a vehicle to collapse due to temperature changes in the trunk. If the fuel carried in the trunk is not in an air-tight container, of course, you will not have this danger, but there will be the danger of the vapors building up in the closed trunk. Both of these conditions can lead to a flash fire in a motor vehicle and should be considered when the fire seems to have started in the trunk.

Friction

Other types of burning in Motor vehicles can include friction materials. All motor vehicles use friction materials which can best be described as materials that are made to rub together. This includes the brakes, clutch and parking brake. While it is not their purpose, these materials generate considerable heat and when not overworked can dissipate this heat. Fire is caused in

FIGURE 36. A broken axle apparently caused the fire that completely consumed this vehicle. The logical explanation for this fire is that friction ignited the brake fluid that was released after the brake drum moved out from the backing plate and the operator kept pumping the brake peddle attempting to stop the vehicle. The area where the fire started was right near the gasoline tank which soon ignited and engulfed the entire vehicle.

these materials when more friction is applied to them at one time than they were designed to handle. An example of this can be a worn clutch which slips when pulling a loaded vehicle up a grade. This slipping is lost motion and there is continual friction on the fiber clutch plate during this time. There may not be an actual fire, depending on the amount of grease and dirt collected in and around the clutch housing but there will be a great amount of smoke. A similar fire can occur when an operator forgets to release the parking brake. In this case also, there may not be a fire but much smoke and the odor of the friction material (brake-lining in this case) burning. This latter case refers to cars with the brake-band around a drum in the driving line under the car, and not the type where the parking brake is connected to the rear wheels.

Carelessness

Occasionally, we have the age-old case where someone will take the gas-tank cap off and then use a match to try to see how much gas he has. Or some joker will try to weld or solder a leak in the gas tank while there is still gas or gas fumes in it. In either case, of course, there will be an explosion or fire or both. This may all seem rather ridiculous to mention but as investigators you will run into these ridiculous causes of fires, and you must not be so hasty yourself that in a rush to find a more complex cause of the fire that you overlook the real cause.

Sometimes you will find an auto burning in a grassy or wooded area. Then you must decide if the auto fire set the woods on fire or the woods fire set the auto on fire. Very often people will burn trash with their auto parked nearby and the fire will get out of hand and burn up their auto.

We had one case where a man was burning trash about ten feet away from his auto which he was working on. Between the fire and his car a can of gasoline was on the ground. The heat from the fire reached the can of gasoline which exploded, setting his auto on fire. The auto was a total loss; the windows were open which allowed the gas to go inside as well as out. It took very little investigation to determine the cause of this fire. We might note that the auto was not insured against fire.

Suspicious Fires

The inside of an auto will not burn as fast if the windows and doors are closed, due to the lack of oxygen. Therefore, a person who is going to set fire to an auto will open the windows or even the doors. So if the weather is cold and you find a burning auto with the windows open this would be reason to be suspicious. Few people will have more than a small vent window open during cold weather.

A check should be made in all motor vehicle fires to insure that all vital parts of the vehicle are present, as sometimes a person will remove vital parts of an auto and then set fire to it and claim insurance on a complete, working auto. Valuable accessories will often be removed before the fire is set. A person who is going to set fire to his own auto or truck will usually remove his registration card and any other important papers from the auto.

If the vehicle is not paid for, check with the finance company and see if the subject is behind in his payments. If he is behind in his payments, he may have decided that it would be more profitable to have a fire and collect insurance than to let his vehicle be repossessed.

Sometimes a check with the repair garage with which the vehicle owner deals may show that his vehicle was in need of repairs that would have cost more than the vehicle is worth. This would be another reason for the owner to have a fire.

TRUCK FIRES

The authors have seen cases where in very cold weather, during a bad snow storm, a tractor-trailer truck would become stranded along the highway and the driver, fearing his load would freeze, would build a fire inside his trailer in an attempt to keep the load warm. And, of course, he not only lost his load but his trailer too. It does not take very much investigation to locate the container in which he built the fire, and usually, upon being confronted with this, the driver will admit to his mistake.

You will no doubt come across trash-truck fires. These, in most cases, start in the trash that is being carried, and are generally due to the operators picking up a load of hot ashes some where along the route. If the truck is of the closed, metal type, most of the time the fire will be contained in this area and not reach the cab. But sometimes the fire department will have to make

FIGURE 37. In this picture we see an auto and a tractor trailer that was involved in an accident. Both vehicles burned. The rear of the auto is badly burned, while the hood and front of the auto was almost untouched by the fire. The trailer and contents were badly burned. Investigation revealed that the gas tank on the auto had burst at the time of the accident. This caused a very hot gas-fed fire, which did a great amount of damage in a very short time. This accident happened on a very heavily traveled highway. This type of fire must be investigated right away, because the highway must be cleared as soon as possible and many of your witnesses will be people traveling through the area and they will stop for a short while and then be on their way. The police who handle the accident can be very helpful to you in your investigation. They will be gathering witnesses for their accident investigation and by working with them you can save much labor in this area. Often the first police on the scene will be able to give you valuable information as to what he saw when he arrived.

the operator dump the load in the street in order for them to put out the fire. To combat this, some trash companies are now putting fittings on top of their truck to which the fire department can connect and flood the body of the truck with water instead of dumping the load in the street.

In farm areas, once in a while you will have a hay-truck fire. This also usually starts in the load. Generally you will have the driver and other witnesses to varify this. Since the hay usually is only on the truck a short time, the most common cause of this type of fire is the cigarette. This may be thrown by accident or on purpose from an auto passing the truck on the highway. This type of fire will very often consume the whole truck.

ACCELERANT CLUES

In the average motor vehicle fire there is little reason for the road under the vehicle to burn. Unless, of course, the vehicle has been involved in an accident and the fuel lines or the gas tank have been broken. It is very seldom that the gas tank of a vehicle will explode due to the fire around it, if the fire department arrives quickly and does a good job. Some vehicles have a Neoprene hose coming out the side of the vehicle with the gas-tank cap on it. In this case, the cap and connecting hose may be blown loose from the vehicle and a stream of burning gas may follow it. If none of the above have taken place and the vehicle is not carrying a flammable load which has spilled onto the road, and the road under the vehicle is actually burning, then we should investigate the possibility that a flammable liquid has been poured onto or into the vehicle and some of it has leaked through to the road. We should also check the gas tank for any holes that may have purposely been made, or any fuel lines that may have been disconnected.

Very often, the rear seat of an auto will burn. This, in most cases, will be caused by a cigarette. Question the operator of the auto as to who was sitting in the rear seat, and whether or not they smoke. If no one was in the rear seat or they do not smoke, then find out if the operator smokes, or if someone sitting in the front seat smokes. Do they have the habit of throwing their cigarettes out the window, instead of using the ash tray? If the rear windows were open and a cigarette was thrown out the front window, it is possible that the air currents carried the cigarette back into the open rear window. This happens rather often.

UPHOLSTERY FIRES

Sometimes when the driver is smoking he may drop a cigarette, or when trying to throw it out the window he drops it in the car. If he does not realize what has happened, or forgets to pick the cigarette up he may have a fire. He may be lucky and the fire will start burning while he is still in his car and he will see it and therefore, the damage will be minor. Or he may park the car and return to find it engulfed in flames. It is possible that no one in the car had been smoking and yet the fire was still caused by a cigarette. If the windows of the car have been open it is possible that a cigarette thrown from another car could have entered this car through the open windows. This may or may not be done accidentally. A cigarette could even be passed through the open window of a parked car by someone walking. In any of these cases it is important to try to find any remains of the cigarette that caused the fire. Then perhaps the brand can be determined and compared with that of the owner and any other occupants of the car. If the brand is different, then we can assume that the cigarette came from outside the car. In a case such as this, if the cigarette was thrown from a passing auto and no one saw it happen, then there is little chance that we will ever locate the offender and find out if it was accidental or incendiary. But at least for insurance purposes we can clear the owner or operator of any responsibility for the fire.

FUSES

When we have an electrical fire in a motor vehicle we should check the fuses. Often people will have trouble with their vehicle blowing fuses so they will wrap one of the blown fuses in silver paper and put it back into the electrical system. In this case, the line may become overloaded and start a fire. This is another case were we may not be able to charge a person with a crime but the insurance company will be interested in our findings.

OUT OF GAS

With the addition of the many expressways and limited-access highways, it is common to find a motorist out of gas. Another passing motorist or friend will often offer to take the stranded motorist to where he can get some gas. There being an extra charge for the proper container from the service station they will decide to put the gas in whatever type receptacle they have or can get which they do not have to return or place a deposit on. After putting this open container full of gasoline in the passenger compartment of their car for transportation to the stranded vehicle, they will naturally announce no smoking due to the heavy fumes in the car and proceed on their way. After arriving safely at their destination, with the fume-laden car, the pair will turn off the ignition and that will be their fatal mistake. Their clothing, reeking with fumes, will make them engulfed as well as the interior of the car. Even if the gasoline was carried in the trunk of the vehicle, the same situation would prevail if it was in an open or improper container. The increasing problem of the motorist out of gas is not only dangerous while transporting the highly flammable liquid to him but also while trying to get it from the receptacle into the gas tank. Then they may find it necessary to pour some into the carburetor to get the car started This, in some cases, is done by the inexperienced with one man pouring the gas into the carburetor while the other attempts to start the car. Here you not only have the possibility of a backfire from the engine but the many sparks from actuating the ignition system including the sparks from the brushes of the starter and generator.

Case Discussion

When a motor vehicle has been in motion and catches fire, you, of course, will have to question the operator. He will tell you whether he was having engine trouble or how the car was acting or what type of mechanical failure he was having. This brings to mind a case where the operator of a motor vehicle that was completely consumed by fire said he was driving along a limited-access highway and tried to slow down. His brakes failed

was quickly extinguished. While heat and smoke damaged the windows of this car, the only fire damage was to the seat cushion.

Upon questioning the owner of this car, it was learned that he parked and locked his car at that location eight hours ago. Since the car was completely sealed when we arrived, we concluded that a cigarette dropped there the previous night apparently started the upholstery smouldering and no one noticed it in the quiet residential neighborhood until the paper boy came along. The owner of this car was not the panicky type and readily agreed that one of his cigarettes tossed out the open window while driving could have come back in one of the open rear windows while he was driving the night before.

An investigator can learn a great deal from small cases like these by noticing the behavior of fire under different conditions. While the lack of good ventilation prevented a complete loss in this case, you can see that with limited oxygen some materials will smoulder for hours or even days. We know that if a cigarette caused this fire it only burned for five or ten minutes itself, then set the upholstery burning. It is not entirely impossible to find the remains of a cigarette, especially in a case of this type. If the cigarette came in the window and landed on the rear seat cushion, it is very likely that the "light" was knocked off it and landed on the rear floor mat or seat cover while the butt itself, being round, rolled down behind the seat on the metal uncovered portion of the floor where there was no fire damage. Upon finding this butt much more information can be gained.

While carrying an investigation to such an extent on a small auto fire may seem like a waste of time, you will find the experience invaluable. When you arrive on the scene of a large fire with much at stake it will be too late to practice.

If you are employed by the department of public safety rapid response to an auto fire will be a requisite. When you receive an auto-fire call, any number of conditions could prevail. There could be people trapped inside, it may be an accident, or other property could be in danger. Even if none of these conditions prevail, the sooner the investigator sees where the fire is burning or appears to have started, the quicker he can conclude his investigation.

SUMMARY

1. In all motor-vehicle fires a complete report will have to be made, as it will be required by insurance companies.

2. Examine parked vehicles for signs of deliberate burning. See if the windows were left open, any accessories are missing (radio, spotlight, etc.) or badly worn tires are on the vehicle. (The good tires could have been removed and sold before the fire and replaced with worn ones.) Check the ground under the vehicle.

3. If the fire occurred in a repair shop, determine what kind of work was being done at the time the fire started. Ascertain if a flammable liquid was being used to clean the engine.

4. A fire in a vehicle-in-motion can easily be investigated by proper questioning of the operator. When he relates any faulty operations of the vehicle an experienced investigator can readily go right to the cause of the fire.

5. An upholstery fire in a closed, parked car is not necessarily an indication that the fire just started. Seat cushions can smolder for hours or days on limited oxygen without extensive burning.

6. Ignition fires may have been caused by using the wrong fuse or wrapping one in tin foil. You can usually find a thermoplastic block under the dash containing all the fuses.

7. Gasoline fires under the hood will have to involve the fuel pump or carburetor and their connecting lines and fittings. These two engine components are the only ones to handle the gasoline before it is sent through the intake manifold into the engine to be fired. Therefore, we can discount the spark plugs, water pump, distributor, starter, generator, etc., as causes of the fire even though their sparks, heat or friction may have set off the fire.

Chapter Ten

AIRCRAFT, SHIP, AND
SMALL-CRAFT FIRES

In some areas this type fire will be more prevalent than others due to its proximity to water, sea ports, air lanes and airports.

SMALL-BOAT FIRES

Starting with small-boat fires, which are probably the most common, we can apply much of the same investigative procedure used in home and motor vehicle fires because, in many cases, small boats are equipped to be motor-powered homes on the water. Power for this type of craft can be put into two classes, the outboard and the inboard motors. The inboard craft with the engine encased in the bilge will create a greater fire potential than the outboard with the motor hung on the transom. During and after fueling is probably the most likely time for a fire to start on an inboard craft. The watertight hull makes a fine receptacle for gasoline vapors to accumulate while fueling, and the first spark from the ignition system can set them off. If the investigator is called to a fire at a fueling dock he can begin his investigation by assuming that the aforementioned has taken place. Fires on small craft that are under way or at anchor can be difficult to investigate because if the fire has any headway, the boat will burn to the water line. Many factors such as paint, varnish, and fuel contribute to this total burning. This is another case where witnesses are the investigators' best recourse. Custom-build or company-manufactured craft generally have greater fire precautions built in than the homemade craft. A homemade craft may not have a marine engine at all, or may have a marine conversion which does not meet safety standards. The fuel for the engine and cooking should be stored so that a leak in the tank will not fill the bilges or cabin with flammable liquid or vapors. Exhaust systems should be water cooled and proper insulation used around the engine compartment. Often, alcohol stoves are used around small craft for cooking and these vapors are just as dangerous as gasoline. When investigating boat fires, the investigator should remember that arson can be responsible as well as an accident. The boat may have cost a great deal of money for which the payments cannot be met, or the hull or engine may need major repairs that the owner cannot afford. Check to find out if the boat is insured then ask around local repair yards for further information as to the condition of the boat or estimates that may have been given. Often a careless boatowner will disregard a small fuel leak and continually have a sludge mixture in the bilges until the time the proper mixture is reached and a spark sets it off. Question friends and family of the owner as to whether such a condition existed.

Anchored Craft

A small craft at anchor that has a fire would almost have to be from a carelessly tossed cigarette or an electrical short. Many anchored craft have anchor lights or use lanterns when at anchor and an improperly secured light can fall to the deck and start a fire. This is usually due to the rocking motion prevalent on anchored craft, whether occupied or not, when a double anchor is used, that is one forward and one aft. This prevents the craft from swinging around into the elements as would be the case of a bow anchor which would prevent much of the rocking. This severe motion of an anchored craft could upset containers of flammables on board that many not be discovered until the next time the owner comes aboard and decides to warm up the engine. The investigator should not overlook the possibility of a fire to cover up a theft on an anchored craft. Check with the owner to determine what valuables and accessories were on board. Robbing anchored craft is a lucrative business in some areas. There are many buyers for the high-priced clocks, barometers, compasses and lights found on well equipped craft if they can be had at reduced prices. The investigator can check with the manufacture or dealer of the missing equipment for a complete description, make and often model which will help in locating it.

Bodies On Board

Of course, if a body is found on a burned boat the investigator takes all the precautions taken in any other sudden death and makes every effort to preserve the scene even if it has to be beached and guarded, at least until he receives a complete report from the medical examiner's office. It is wise to preserve the scene until this time because determining the cause of death can alter the course of the investigation and give you some idea of what type of clues to seek. At any rate, if arson of any kind is suspected, the investigator should remember that a burned boat is harder to preserve than area on land and he should act accordingly. Among the vital information that can be obtained from an autopsy report in reference to fires and death on land, sea or in the air it can be determined if the deceased was alive during the fire and if he inhaled any of the smoke or fire. It is not likely that a person who was dead before a fire would have scorched wind pipes. Also, the time of death can be compared with the time of the fire. Burned hands, arms and face may indicate the deceased attempted to extinguish the fire. If a body has been severely burned on the outside and the medical examiner reports no sign of internal residue from the fire you can assume the subject was dead before the fire began. This does not necessarily mean a homicide, as the subject could have been stricken while cooking, smoking or even striking a match. Again you would have to depend on the autopsy to determine the cause of death, which, unfortunately, sometimes is never learned. Investigators should educate themselves in recognizing possible causes of death from the appearance of the deceased. It is also possible for the investigator to approximate the time of death and to determine if the body was moved after death. This information is covered in Chapter Fourteen and coupled with any other information you can procure will speed up any fire investigation where there is a death involved. If you suspect a body was placed in a small craft after death, you can assume it was done with some difficulty due to the compactness of the craft and most likely had to be dragged to the place where it was found. Remember in the case of a death and fire unless you are certain it is accidental do not move the body without taking proper photographs and making notes of everything in and around the craft; you will never be able to duplicate the setting once it is disturbed.

AIRCRAFT FIRES

In discussing aircraft fires we will divide them into two classes, the small private planes and the larger usually commercial type planes. Military aircraft will fall under either of these classes and will not be discussed separately. Most plane fires will involve a plane that has crashed, so we will cover this area first. As in all fires, punctuality in arriving at the scene is a must for the investigator. The immediate area of the crash must be closed to all unauthorized persons and a large area surrounding the crash scene roped off. Souvenir hunters and the curious will converge on a plane crash scene and become a primary hindrence to the investigator. All persons seen with objects connected with the crash in any way should be detained. These persons will have to be questioned as to where they found the objects and checked for possible personal belongings of the victims. Assistance of other police on the scene can be enlisted for this task.

Upon arriving at the crash site the first thing to determine is whether the crash was a result of the fire or the fire was a result of the crash. In the large commercial aircraft crashes, there is almost always fire. If there was fire before the crash, it will spread after the crash. Due to the spreading of high-octane gasoline total burning is not uncommon. Nevertheless there are many factors to be considered in determining if the plane was on fire before the crash. As in all fires, one of our best sources of information is the eyewitness. Unfortunately, in this type of catastrophe there will be many unreliable witnesses either because they are panicky or just want to get into the act. You will have to evaluate all the statements and determine which ones fit together. A lot can be learned from close examination of the victims at the scene and from the autopsy report. In examining the wreckage, try to determine if any automatic or manual firefighting equipment was in use while the plane was in flight. The position and location of the victims can give a lot of information to the investigator regarding how much, if any, warning they had before the crash.

The pilot and crew of a doomed aircraft will do everything possible to protect the passengers and prepare them for the crash. The passengers will be tightly strapped in their seats and may have their heads in their laps and pillows tucked around them. If wreckage of the plane is spread over a wide area (many miles), it either exploded or came apart in flight. Wreckage spread over a large area at the crash scene does not necessarily indicate this, as a plane traveling hundreds of miles an hour striking a solid object can result in particles being strewn over a wide area. Some sort of diagram or map of the area should be obtained or drawn by the investigator and the location and description of anything relevant to the crash noted on the map including locations of where witnesses were when they saw what they said in their statement. Parts of the plane and personal belongings of the passengers should all be located on the map and measured with some relation to a fixed object on the map (a house or road or intersection). There will be many other investigators at the scene and you will have to cooperate with them. Eventually all the wreckage will be gathered and reassembled or at least parts will be put as near to their original position as possible. Examining the reassembled wreckage along with the diagram indicating where each part was found will be instrumental in determining just what series of events took place just before the crash. More information as to the cause of the crash may be contained on a tape in a sealed fire-proof container which commercial aircraft carry. Depending on whom you are employed by, the recovery of this tape may not come under your jurisdiction and should be left for the proper authorities. When your investigation indicates the plane came apart in the air or exploded while in flight, do not overlook the possibility of a bomb having been placed on board. Careful examination of the baggage compartment should be made to determine if there has been any outward pressure on the bulkheads (walls) of the compartment particularly between the passenger compartment. If you locate any compartment with the bulkhead appearing to have been blown out or into other compartments, it is reasonable to assume an explosion occurred in that compartment. Fragments of the bomb may be located embedded anywhere in the compartment as well as parts of the suitcase or other container that held the bomb. Of course, there could have been an accidental explosion also, but it would not likely occur in the baggage compartment. If you are a local investigator and discover evidence of a bomb or have any suspicion of foul play in connection with a plane crash, you should contact federal authorities immediately, as it would be a federal offense.

Fires in aircraft can have causes much the same as other types of fires. Gasoline vapors, leaky fuel lines and electrical short-circuits can cause a plane to catch fire while in flight. Lightning or electricity has been, from time to time, suggested as a possible cause of an air disaster but this is difficult to prove. In the case of the small private aircraft, engine failure would be a more common cause of the crash or human error on the part of a pilot. A small plane pilot may become lost and run out of fuel or try to make a blind landing without proper equipment. There should be little difficulty for the investigator to determine if the fire of a crashed plane was fed by an accelerant, which would indicate, of course, whether there was fuel in the tanks. Check the propeller of any downed craft to help you determine whether the engines were operating when the plane hit the ground. Examine the blades: if those which are not near the ground are bent on the tips, they were rotating at the time of the crash; conversely, if only the blades nearest the ground are bent, it is reasonable to assume the prop was not spinning at the time of the crash.

In the case of a small plane crash there will be an official investigation, of course, but due to the comparatively few people involved, the investigation will not begin to compare with the investigation of a large commercial plane crash in which many people may have been killed while traveling on a licensed public carrier. You will find that many of these investigators are learned men who specialize in plane crashes. Even with all this knowledge at the scene of a plane crash, many times the cause is never definitly determined. Adding to the difficulty is the fact that the best witnesses, who could give the most valuable information about a crash, were the crew of the downed plane which had no survivors. It is not uncommon for a downed plane to be so disintegrated that it is impractical to gather the pieces and attempt to reassemble them for study. In this case, it is best to have an overall diagram of the crash scene identifying the location of each part, and then to gather the parts for examination.

We have discussed lightning as a cause of fire on the ground and suggest the possibility of it being involved in a plane fire or crash. Even if air turbulence, which accompanies violent electrical storms, caused structural stress that was responsible for fire or crash, we can consider lightning as being involved. Severe plane crashes during lightning storms are a matter of record, and there are witness statements regarding planes being struck by lightning. Lightning, however, is considered uncertain as a *cause of fire* in aircraft, as is static electricity. Static electricity can be built up in fuel tanks while fueling and can also be created by the motion of the fuel in the tanks during severe turbulence. While static electricity is dissipated from motor vehicles by means of a ground strap, aircraft discharge it off the trailing edge of the wings.

It is always important to determine the exact time of a fire and a plane crash is no exception. There are many ways this can be done. If the plane crash knocked out electrical wires, some home in the area will have an electric clock that stopped at the exact time of the crash. Watches of the victims can also be examined as well as checking the time the first alarm was sounded by an eyewitness. The radio log kept by a control facility which had contact with the downed plane will also have valuable information.

SHIP FIRES

Navy Ships

It is not without reason that the Navy puts great emphasis on oil fires. With a ship made of steel and someone on watch all the time it is not likely any other kind of fire could gain much headway. All ships have fire stations and equipment throughout and men assigned to man these stations. So aware are men of ships of the dangers of oil fires that while fueling at sea or in port it is required that a red flag known as "Baker" be flown from the mast. This red flag notifies all other ships and persons in the area that fueling, handling dangerous cargo or some other delicate operation is taking place aboard. An investigator arriving at a ship fire who notices "Baker" flying will know that a dangerous operation was in progress.

Tankers

In the case of a tanker which is or has been loaded with flammables and which catches fire while in port, witnesses may be difficult to find. We have seen tanker fires of which it was inpossible to get within hundreds of feet, due to the heat and flame. This makes investigation difficult but it can be started at the office of the port. There you can determine why the ship was in port, what repairs were to be made, and a record of faulty equipment.

While the exact cause of a fire of this type may never be determined due to the complete burning and the absence of witnesses, a probable cause can be entered in your report derived from the information you received from the port office, and the owners of the vessel. Examining the burned hulk after it cools off may add to your information if you can locate charred ruins of a burning torch that was in use (which may be determined by the position of the valve) or other tools in an area where repairs were scheduled to be made.

Passenger Ships

A fire on board a passenger ship can often produce as brilliant a fire as one on a tanker or an inflammable-laden freighter. The efforts of the ship's owners to please the passengers will lead them to install highly finished paneling with elaborate dining rooms and bars. This is all fuel for the fire. If passengers are on board at the time a fire starts, the investigator should consider the festive mood that prevails on most passenger liners. A fire in a passenger's cabin is more likely to be an indication of a careless passenger than negligence of the crew or owners of the vessel.

With the many cabins, compartments and corridors on a passenger liner, an experienced investigator should have little trouble tracing the course of the fire and determining the approximate area in which it started.

Freighters

A cargo ship, except for the aforementioned fueling hazard is relatively safe. The fire danger lies in the type of cargo it carries and the way it is loaded and stowed. Friction and spontaneous combustion can be the causes of fires in improperly stowed cargos.

Dust from materials being poured into a hold of a ship can cause an explosion. Many substances that are not highly combustable in bulk form will be highly explosive when atomized. The business of knowing how to stow cargo and how to handle it is a profession in itself and an investigator confronted with a fire or an explosion in the cargo hold of a freighter or while being loaded or unloaded at a pier would do well to gain the confidence of one of these professionals. An experienced dock worker will more than likely be able to give some helpful information as soon as he learns the type of cargo being handled. Never hesitate to question men in a particular trade where a fire was involved. Most tradesmen are fully aware of the particular dangers of their trade and often can make a suggestion that would have taken days to learn if you had not consulted them.

SUMMARY

1. In all instances where fueling was in progress when a fire started, it is safe to commence the investigation with this process in prime consideration.

2. An explosion and fire at the instant the engine is started will indicate a fuel-line leak or oil-soaked bilges. This type of fire often occurs just after fueling, while the gasoline vapors are still in the craft. The initial spark from the ignition is all that is needed to set it off.

3. In cases of large catastrophes, such as in air crashes, the investigator will have to sort out the genuine witnesses from the publicity hounds. The actual witness may be reluctant to offer information for fear of publicity.

4. Investigating a fire in the cargo hold of a merchant vessel can be speeded by consulting persons familiar with the type of cargo and the accepted methods of storing it.

5. The gala festivities on a passenger cruise ship that were suddenly dampened by a fire can lend much confusion to the investigation. The best source of information would be the crew rather than the passengers.

6. Tankers, which often totally burn, are difficult to investigate. However, the ship's owners, authorities at repair yards and seamen who were not on board at the time of the fire can all be of help to you.

7. In propeller-type plane crashes, it can be determined with reasonable certainty if the engines were in operation at the time of the crash by observing the condition of the propellers.

8. If sabotage is suspected in an air disaster, "character probing" of the victims will be necessary. This will include excessive insurance policies purchased recently. Also persons who missed the flight will have to be questioned.

Chapter Eleven

FIELD, BRUSH AND WOODS FIRES

IT is probably very seldom that any of these fires will start by natural causes. The most usual of the natural causes are lightning and "natural" combustion. In most cases, man is the cause of the fires whether they be accidental or incendiary. Fields and woods naturally burn more easily during the dry months when they are not too green. This is generally during the spring and fall months. But any time we have dry weather the fire danger increases. We must also consider the fact that during the warmer months there are more people in the fields and woods, including the children who are off from school, therefore, the chances of woods fires of any type are greater. In the fall we have the "hunter problem."

How Fields Burn

Let us first consider the field fires. The dryer the field, the faster it will burn. The fire will burn in the direction the wind is blowing until it hits a stream or cleared area. But depending on the size of the cleared area and the power of the wind, the fire may cross the clearing. The fire will usually burn very little against the wind. If there is no wind and the fuel permits, the fire will burn in a circle working from the center (point of origin) out. Unless the wind does not permit it, the fire will burn faster up hill than down. Of course, some grasses, such as sage, will burn faster than others.

How Woods Burn

Woods, which are dry and not too green, will burn fastest. The thicker the underbrush and accumulation of leaves or pine needles on the ground, the faster the fire will progress. Here again, the wind permitting, the fire will tend to burn in a circle starting from the point of origin and working out. If there is a wind, the fire will burn with the wind. The fire will generally travel faster up hill than down. If the wind is right it is not uncommon for the fire to get into the tree tops and travel from tree top to tree top at a very high rate of speed. This is especially true in pines. This tree-top fire will travel at a high rate of speed often leaving the ground-level fire far behind it and, in many cases, trapping fire fighters. On the other hand, if there is a heavy mat of pine needles on the ground it is possible that the fire will smolder under the surface out of sight and break out when and where it is not expected. The fire fighters may think the fire is out, but when they leave it will start up again.

The authors feel that all field and woods fires should be investigated as they are a great loss to the country and there is a great expense in fighting them. A check with any rural fire department will show that a great number of man hours are spent fighting field and woods fires each year.

INITIAL INVESTIGATION

The investigator, as in all fires, should arrive on the scene of the fire as soon as possible. On his arrival he should note the time, wind direction, extent of the fire and the people on the scene. Then he should report to the fire-fighting officer in charge and find out the time they got the call, the time they arrived on the scene, and the wind direction and extent of the fire on their arrival. A check with fire department headquarters may reveal the name and address of the person who reported the fire. If possible, the person who reported the fire should be questioned as to the time they first saw the fire, the color of the smoke, size of the fire, wind direction and any persons or auto they may have seen in the area. All persons in the area of the fire or in homes overlooking the fire area should be questioned. From information received from the people questioned and your knowledge of how field and woods burn, you should be able to determine fairly closely the area where the fire started.

FIGURE 39. There is little question but that the house shown here is a total loss. It appears that there is a slight breeze blowing to the right of the picture. There is another fire burning behind the metal shed shown in the right side of the picture. A fire of this nature is a great danger to other buildings nearby. Sparks from a fire like this will often travel a great distance, if the wind picks up. These sparks will set fields, woods, and other buildings on fire. The investigator must keep this in mind if he gets several small fires in the area of a large fire.

Examining the Scene

This area must be examined very closely in an attempt to locate the exact point where the fire started and any evidence that may be on the scene. We will look for stained and heavily burned areas where possible liquid flammables were used to start the fire. The remains of matches or match packs are always good clues. It is always a must to make plaster casts of any footprints found in the area. Investigation may show that there were actually several fires, not one, and this, of course, kills the idea that the fire may have been accidental. It is possible that when checking the fire area we will find several stones placed in a circle which may indicate the location of a campfire; most campers will put some stones around their fire even if they do not properly clear the area. This, unless other evidence is present, may indicate the fire was started accidentally by campers who failed to use good safety measures. Often the point of origin will be directly beside a road, in which case, it may be decided that a cigarette was thrown from an auto. But we must not jump to this conclusion, as this may not always be the case. In fact even if it is the case, it may not have been done accidentally.

We will sometimes find that the woods or field fire started when a small shack, lean-to, or some other type of camp building caught fire. Investigation will tell you whether this was a child's play house or the home of a hobo. Most of the time you will find it the case of children playing with matches and carelessly setting their shack on fire. It usually will take very little investigation to find out which children use the shack.

Lightning

If there has been a storm recently, lightning can be the cause of a woods fire. Sometimes the lightning will strike a tree and cause it to burn. Often just the tree that was struck will burn but, in some cases, the fire will spread and start the whole woods burning. The authors have seen a case where just the top of one tree in the middle of a woods was burning—an oak tree about three feet in diameter and approximately seventy feet tall.

Review Fire Records

Often, even if we find no direct evidence as to whether the fire was accidental or incendiary, a check of fire records may show that there has been a large number of fires in this area. This shows us that the fire bears further investigation and may

FIGURE 40. In this picture we see large piles of sawdust burning. This fire not only involved the piles of sawdust but also a building and several trucks that were at this location. The smoke is blowing to the right and away from the picture. The pile of sawdust behind the firefighters is partly burned and is slowly smouldering towards the firemen. This pile will smoulder and burn for a long time if left alone. Large amounts of water under pressure was necessary to put this fire out. Any area where there is large piles of sawdust like these can be a trouble area because the local children will consider this to be a good place to play.

be the work of juvenile fire setters or a pyromaniac. By comparing all of the fire reports, we should be able to determine a pattern. If you find the fires are occurring after school hours or on the week-end (if school is in session), or if you find that the fires have also occurred on one weekday which was a school holiday, this would definitely indicate a school-age juvenile fire-setter. When fires are being set during the school hours a check with the school will show who, if anyone, was absent on these days.

When this has been investigated and the results are negative, you can then assume that your fire setter is above or below school age or has quit school. The school can often be helpful in finding out who in the area has quit or been expelled from school. We must not rule out the children below school age as being too young to set fires, as it is a known fact that some very young children will set fires. This is especially true of field and woods fires. Naturally you should make a check of your records for any known fire setters who may be living in the area. Since, by this time, you have established a pattern to the fire settings, you may find it necessary to stake out the areas where the fires have been occurring on the days when they are most likely to happen.

Accidental Causes

One of the most common causes of field or woods fires is the person burning trash, leaves, or other debris. There usually is not much investigation to do in a case of this type. In most cases, it is easy to determine what has happened and often the one who was doing the burning will call the fire department and admit what has happened on their arrival. This type of fire can then be ruled as accidental, although you may wish to place charges for violation of local burning laws. If the subject does not admit to his error, by questioning a few neighbors you can generally clear up the case very quickly.

In areas where steam-driven locomotives are still used it will be found that often they will start fires along the railroad tracks. The diesel engines have greatly reduced this fire problem but have not completely done away with it. There is, of course, the possibility that someone can throw a cigarette from a train and start a fire. Railroads are favorite playgrounds for children, therefore, we cannot discount the fact that they may start a fire in this area. In fact, they are very likely to be one of the major causes of field and woods fires along the railroad. We must also consider that we do have hobos and bums traveling and camping along the railroads and, in many cases, they may be the cause of fires in the area. Often the railroad police can be of assistance to you in cases of this type.

Sabotage

Because our forests are so very important to the welfare of the country, we must remember that the forest fire can be an act of sabotage. The saboteur might use a timing device in setting his fire. The device could be as simple as a lit cigarette in a pack of matches. When the cigarette burns down, it sets off the pack of matches thereby setting afire the surrounding timber. Naturally, if sabotage is suspected, we will notify the federal authorities and gain their assistance in the investigation.

No matter what type of field or woods fire we are investigating, we will find we are not the only ones interested. The local fire department, state forestry wardens and the state game wardens will, in most cases, be very interested and will assist you in any way possible.

Hunting Season Danger

In the fall months, when the woods are dry, we have an increased fire danger due to the large number of hunters in the woods. Here we have a large number of people wandering through the woods, many of whom are not trained in the ways of the woods. They will, in some cases, be building fires for cooking and keeping warm. And, of course, many of the hunters will be smoking. Because of this we are sure to have some accidental fires in the area.

Improper Extinguishment

When a forest fire breaks out in an area where there has recently been a fire, there is the possibility that even though the fire appeared to be out, it could have been burning underground. A fire can smoulder in the roots of trees for days where there has been a forest fire without being noticed and then break through to some air and start the forest fire all over again. Changes in the weather to permit more favorable burning conditions will have a lot to do with a smouldering fire, especially if the air clears and the wind increases. This will dry out the already extinguished timbers and give the new fire the air it needs to get a good start.

All damage done by this second fire can be attributed to the same cause as the first fire, as it is one continuous fire and the fire fighters could not be expected to dig up all the roots in the burned area to see if they are burning underground.

SUMMARY

1. Most field and woods fires do not start of natural causes.
2. The fire will travel faster in the direction the wind is blowing. It will burn faster uphill than downhill.
3. It is possible that in a woods, the fire will travel at high speeds through the tops of the trees. Or if there is a heavy mat of pine needles on the ground it may travel under the surface, out of sight.
4. On arrival at the fire, the investigator should note the time of his arrival, wind direction, strength of the wind and extent of fire. Then compare this with the same information received from the first fire-fighting officer who arrived on the scene.
5. Question all people in the area.
6. Pinpoint the area where the fire started and examine it closely.
7. Look for physical evidence such as match packs, signs of a camp, discharged shot-gun shells or timing devices.
8. If there has been a lightning storm, look for signs such as a damaged tree which could indicate lightning struck.
9. Review fire reports on other fires in the area.
10. Consider the possibility of sabotage.
11. Check to determine if there has been a recent fire at the same location which may not have been properly extinguished.
12. Remember most field and woods fires are caused by humans, be it accidental or incendiary.

Chapter Twelve

ELECTRICAL FIRES

W̵E will go into electrical fires at great length because this is one of the primary causes of accidental fires. Despite the large number of fires which can be traced to electricity, the acceptance of electricity has greatly reduced fires that were formerly caused by other means of lighting such as candles, oil lamps, gas lights, etc. It is usually improper use and installation of electricity that makes it dangerous. An investigator suspecting electricity as a possible cause of an accidental fire can observe the wiring in any section of the burned building and know if it is adequate and safe.

FUSE BOX

A wealth of information can be obtained from the fuse box. The fuse box is the terminal point for each circuit in the building and each fuse in the box will service a complete circuit. A circuit is one complete path of electricity, all the outlets connected to that one line and those serviced by the fuse on that line. The investigator can count the fuses in a fuse box and determine if the house has adequate wiring. As an example, a small home may be sufficiently supplied with only four circuits but larger homes or building would need six, eight or possibly more to be adequate. In some areas this is controlled by state or local laws. Even with rigid laws it is impossible to inspect every circuit, outlet and appliance in every home or building. The investigator should take a picture of the fuse box for future reference. Before the investigator goes any further he should make sure the main switch is off. This may be done by pulling a switch on the side of the box or removing cartridge fuses. (If in doubt consult the utility company.) You may notice fuses of different

FIGURE 41. This actual photo of a fuse box in the cellar of a house on fire gives a clear indication that one of the fuses is blown out. (The one on the left.) This is a four-circuit fuse box which is all right for a small house and the area served by each fuse is indicated on a diagram on the door. If no diagram for the fuse locations can be found you can open the entire front of the fuse box which will expose the wire connections to the fuse sockets and then you will know what wire leads to the area served by the burned out fuse. At the taking of this picture the main line to the house had not as yet been disconnected and the investigators standing in a few inches of water decided not to make any further examination of the fuse box at this time. When examining a fuse box make a note of the ratings of each fuse in their respective sockets as you may need this for future reference and someone may change the fuses after you leave. This particular fire was found to have been started near a wall receptical serviced by the darkened fuse which had an appliance plugged into it with an admittedly faulty line cord. There were no injuries but damage to the house was considerable.

FIGURE 42. This is a picture of a large farm house. The house is of frame construction with wooden shingles. As its plain to see this fire took place during the winter. By looking through the hole in the wall near the right side of the picture you can see the remains of a partition between two rooms. It was in this partition that the fire started. It is believed that worn wiring was the cause of this fire. It was necessary for the people who were in the home to run several hundred yards to get to a telephone to call for help. With this delay and the fact that there were no fire hydrants in the area, it was only through the good work of the fire department that the home was saved.

FIGURE 43. In this picture we are looking at the dormer which was part of an apartment on the second floor of this home. The fire started on the second floor and was very extensive in the area of this dormer. A close examination of the windows in the dormer will show you that the wood between the right and center windows is almost burned through. Looking through these windows you can see that this room was badly burned. The fire was confined to the second floor. This fire was accidental.

rating or even open (burned out) fuses. Some types of fuses will indicate whether they opened because of a direct short or an overload. When you locate an open fuse, unscrew it and look in the socket and you may find the cause of the fire. The most common object found in a fuse socket is a copper penny. This is done when the circuit is overloaded and keeps blowing fuses, a penny may be inserted into the socket to complete the circuit. At this point we will mention that an overloaded circuit will break down (burn) at its weakest point and after the penny is in the fuse socket, the fuse offers no protection and a fire may break out anywhere along the circuit which cannot handle the extra load. This jumper in the fuse socket may have worked for days or months until the insulation around the next weakest point breaks down. In many older buildings the investigator can find one or two outlets in the entire building with multi outlet sockets added and wires running in all directions. This is another example of inadequate wiring and a reason why people put pennies in fuse boxes.

TRACING CIRCUITS

Invariably when there is an electrical fire there is some evidence of it in the fuse box and the investigator can follow the wires from the suspected fuse and confirm if they lead to the area where he has already determined the fire may have begun. Even if the circuit itself was not shorted or overloaded it may lead to an appliance or fixture which was the cause of the fire. It is even possible for a fixture or appliance to start a fire without being faulty and, therefore, not show up in the fuse box. This would include an iron or toaster situated too close to flammable material (or even a light bulb). Any appliance or fixture which produces heat can cause fire through carelessness. This would not be an "electrical" fire and it is not likely any clues would be found in the circuits. When you have traced a suspected wire to a definite area, start from the nearest outlet box. There may have been an extension wire tacked along the baseboard that began to burn and started the fire. Remnants of the copper wire may be found on the floor. These extension wires are more likely to cause fire than the heavy cable that is run from the fuse box. The longer the wire is and the more watts it handles, the more likely it is to burn. Also determine the position of the "off-on" switches on all equipment on the suspected circuit. Adding up the wattage ratings to all the equipment in use on one circuit at the time of the fire can help you pinpoint the origin. Also remember that electric motors produce a slight overload when they are started. It is generally acceptable for a circuit having electric motors on it to have a heavier fuse but this does not compensate for the flimsy wire often used as an extension. The cable to the fuse box will often stand an extra load but close examination of the two wires inside the armor that are connected in the fuse box will often reveal some discoloration from heat. The suspected wires can be compared with other wires in the fuse box. In the case of a building that has been properly and adequately wired, a direct short circuit will blow the fuse immediately and will not usually leave the discolored or brittle insulation that an overload will. However, wherever the short occurred, the flash may have been sufficient to start a fire. In most wired buildings all terminal connections are made in junction or outlet boxes. These boxes are reasonably fire proof and safe if good connections are made; however, it is not unusual for vibration to bare a wire in a box and cause a short that will result in a fire inside the wall especially if the fuse does not open immediately.

Related Causes

Since some wood will ignite after a half hour of exposure to 350 to 400 degrees F. you can see that a light bulb, iron, heated wire or any device that produces heat can start surrounding material burning. In investigating electrical fires the investigator can often be misled by the apparent origin being located in more than one place giving rise to the suspicion of arson. However, the electricity travels all through the walls of the house and especially where lightning is involved the burning may appear at two or more places. It is not uncommon in electrical fires to detect an odor similar to that left by lightning, especially if there

FIGURE 44. This building, as you can see, was a rather old and of frame construction. This home was located on a farm in an area where there was no water supply. The large amount of white smoke seen in the upper part of the picture is caused by the hose line which the fireman is using on the roof. They are using a fog spray to cool and smother the fire. This is actually a sign that they are getting the fire under control. This fire started in a room located under the roof where the firemen are standing. This room had been paneled with plywood and the plywood was completely burned away. This fire was started by a short circuit in the electrical system, and a brilliant glow in the sky was apparent for miles.

was a violent short causing prolonged arcing and fusing together of metals. Witnesses often tell of seeing a brilliant flash or of actually hearing the short. The odor left by the arcing of electricity into the atmosphere will more likely be familiar to the first firemen on the scene and they will be the best source of information on this.

OVERLOADS

In an inadequately wired house where a fuse is blown the investigator can check the wall recepticles in the area where the fire started. It is not uncommon for people with inadequate circuits to plug two triple plugs into one double recepticle making a total of six outlets and then indiscriminately plugging equipment into these outlets that were only made to handle two plugs. If you look around the floor where the fire started you may find the remnants of these multiple plugs which you will easily be able to identify.

It is often sufficient to conclude that a fire started in a certain appliance without definitely pinpointing the cause. You could say a fire started in a washer, dryer, toaster, iron or any other appliance, or in an electric motor used to operate any of these or other appliances, equipment or tools. However, to make a really thorough investigation you should determine what caused the particular piece of equipment to start the fire.

ELECTRIC MOTOR FIRES

In most cases where an electric motor is part of the equipment it will be noticed that the motor is completely charred inside. However, this does not mean that the motor was faulty. In the case of a washer where the motor burned, it may be traced to an overloaded washer or a foreign object in the water pump that jammed it and prevented the motor from operating when power was applied. If the motor was able to operate with a part of the washer jammed, the fire could have been started from friction of the belts or pulleys. In cases similar to this it is not fair to the manufacturer of the motor for you to put in your report that the motor caused the fire, when it was actually caused by improper use of the appliance. If a fire is traced to an electric motor and the rest of the equipment that it was operating is found to be in good working order, you may find the motor had worn bearings that were not lubricated, faulty starting mechanism or excessive dirt in the motor. Excessive dirt in a motor is often found where the motor was used to power a saw. Refrigerator fires are not common because no heavy load is placed on the motor, but if one does occur it can usually be traced to the starting mechanism. Most people will have a heavier fuse in the circuit that bears an electric motor and will be able to tell you that the appliance made an unusual noise or the lights went dim before the fire broke out. This is a clue that the starting mechanism in the motor failed. Other electrical equipment such as lamps, irons, toasters, etc., that are suspected as being the origin of a fire more than likely had worn or damaged line cords or faulty connections inside the appliance. This same reasoning applies to radio and television equipment which may have a part burned inside but is well protected against the fire getting outside of the chassis.

SUMMARY

1. In accidental electrical fires, much information can be gathered by inspecting the fuse box. Discolored wires, oversized fuses, blown or "open" fuses and new fuses that have just been installed are all indications of previous electrical trouble.

2. A blown "open" fuse in a fuse box does not always indicate a dead circuit. Pull the main switch and see if there is a penny or other metal object in the socket with the fuse.

3. A circuit diagram usually can be found on the inside of the door of the fuse box. This will help you identify the fuse you suspect and tell what section of the building it services.

4. If you are unable to trace a circuit because it disappears into the walls and the wiring diagram is illegible, consult the respective utility company in your area. They have equipment to trace a circuit and are very cooperative.

5. Examine the recepticles and outlet boxes in the room to which the suspected circuit leads. The metal prongs of multiple-outlet plugs may still be protruding from the socket. The remains of an extension cord may be on the floor in the form of a thin blackened wire.

6. Check the position of the switches on the remnants of all suspected electrical appliances. However, if the line cord or plug to an appliance was damaged or worn and caused the fire, the position of the switch (on or off) located on the appliance may not be important. This would depend on whether the wattage drawn through the wire and plug by the appliance in operation caused the breakdown.

7. Violent electrical arcing, as well as burning electric motors, transformers and insulation all give off a distinct odor. An investigator should question witnesses and firemen to this effect.

Chapter Thirteen

FALSE ALARMS

IT is important that we also talk about false alarms because, in many cases, the same men who are required to investigate fires are also given the responsibility of investigating false alarms. Most of us are familiar with the cost, danger and other problems presented by false alarms so there is little need to spend much time here on this. I feel we all agree that false alarms are a great waste and, therefore, it is important that they be investigated.

BOX ALARM

One of the most common types of false alarms is the type given on a fire alarm box where some type of door is opened or glass is broken, and a lever is pulled, which transmits the box number to the fire department. In this case, the person giving the alarm has no direct contact with the fire department personnel, therefore, this method is very popular with the type of person who wishes to turn in a false alarm.

Who Will Pull A False Alarm

You may now wonder just what type of person will pull a false alarm—just about any sex or age of person has been known to pull boxes on false alarms. Children so small that they have to stand on their three-wheel bike to reach the lever have been known to turn in a false alarm. A female who has a boy friend or husband in the fire department may pull a box to see her man ride the fire engine. Here again, a man rejected from the fire department may pull a box in order to get some action. Or some drunk may pull a box in the middle of the night because "those bums down at the fire house are getting paid so they ought to be awake." Often, teenagers will pull one or a series of boxes just for kicks. In some cases, boxes will be pulled to busy the police elsewhere while a crime is being committed. These are just a few of the popular reasons for turning in false alarms. No doubt just about any reason you could think of has been used. We must realize that although many of these people may only pull a box one time or a few times and then never pull another box in their life, for some people this will just be a start. Some will pull the box for the personal thrill it gives them to watch the fire engines arriving on the scene. They may pull many boxes and soon find it is losing its thrill and then decide if there was a real fire it would be more thrilling. Then you have a fire setter on your hands.

Investigation

Naturally when a box is pulled it is important to question anyone living or working in the area as they may have seen who pulled the box. Or they may be able to give a description of anyone they saw hanging around the area or any suspicious autos in the area. The day and time the box is pulled can be helpful. In most areas, on a weekday most of the men are at work, this cuts down the number of men you can suspect. If the alarm is pulled in the middle of the night, in most cases, this would eliminate the young child standing on his three-wheel bike to reach the lever. Of course, in a case where more than one box is pulled in the area, then it is easier to determine a pattern as to when or where the next box may be pulled or who may have pulled it. In many cases, we will find that we will have a whole series of boxes pulled in one night. In this case, the exact time each box is received at the fire department is very important. By comparing the times between alarms and the distances between boxes, it can often be determined if the subject pulling the boxes is walking, riding a bike or an auto. In many cases, an auto will be used for this type of offense.

When you are having trouble in one area it is possible to coat the handles of the boxes with invisible dye that cannot be seen until placed under a black light. When a subject pulls a box he will then have this dye on his hands.

Close cooperation between the fire department and police department is very important in any cases where boxes are being pulled. The police department should be notified of the box alarm as soon as it is received by the fire department. In many cases, there may be a police car very near the box when it is pulled, and if they are notified right away, they stand a chance of arriving on the scene while the subject is still in the area.

TELEPHONE FALSE ALARMS

The telephone is another method used for turning in false alarms. The same type person who will pull a box may telephone a false alarm. But, in most cases, an individual will stick to one method of turning in an alarm. Here again age or sex of the person may vary. Young children playing with a telephone may call the fire department and report a fire. There have been cases where teenage girls while having a party would call the fire department and report false fires just for kicks. Some people will call the fire department in the middle of the night and send them to some person's house that they have a grudge against. Often the address given the fire department in telephone false alarms will turn out to be that of a vacant lot.

Use Of Tape Recorders

Many fire departments now have tape recorders hooked to their telephones so that all phone calls are recorded. In cases of false alarms, this is very helpful as it gives us a recording of the voice of the person wanted. We can play this recording back and determine the sex of the subject and listen for an accent or any speech defects. It is also important to listen for any background noises, as they may help to determine from where the call came. When more than one false alarm of this type has been received, the recordings can be compared to see if the same person, or more than one person, is making the calls. If past recordings of false alarms are kept on file we can go back and compare with recordings used in cases cleared and we may find a subject that we have convicted in the past is out and back to work.

Investigation

The area around the address given should be checked for suspicious people or autos and all residents of the area should be questioned. Very often, public pay phones are used for reporting false alarms, so it is important to note all pay phones in the area and to question residents around these phones and also businessmen who have pay phones in their stores. The police who work the area should then keep an eye on all pay phones in the area. They should note the description of any persons around pay phones and the exact time they saw them. They should also note the tag numbers of any auto parked near a pay phone. The times on these notes could then be compared with the time a false alarm is received and may prove very helpful.

No matter what method is used to turn in the false alarm, it is not usually any easy job to arrest the offender. But the more alarms he turns in the greater the odds are in your favor. Through good investigation, patience and a little luck, in most cases, you can arrest the person pulling the lever or making the phone call.

Case Example

In some cases, you can catch the person who has pulled an alarm for the first time. Such was the case where a pre-teen boy pulled a box alarm about a block from his home and then ran home. No doubt he gave the fact that there was snow on the ground very little thought; but this made it very easy for the investigator to follow his footsteps which led directly to his home. And, of course, a comparison of his wet boots with the footprints at the scene revealed they were the same size and had the same sole and heel patterns. It is, of course, important in a case such as this to take photographs of the footprints before the snow can melt and destroy your evidence.

Evidence

Even if there is no snow on the ground it is still important to check the area around the box for footprints. If there is soft ground in the area we may find good footprints that we can measure, photograph and then make a plaster cast of. In many cases, there will be no snow or soft ground around the box, but there will be loose cinders from the road. This is not good for footprints but it is possible to find a rough footprint good enough to be measured and give you a general idea as to the size of the subject's foot. Sometimes at night when there is a heavy dew on the grass, a wet footprint may be left on the sidewalk. Any information that can be obtained from footprints is very important. The size of print will give you a general idea of the size and age of the person for whom you are looking. But lets not forget that some teenagers today have feet the size of a grown man. Naturally, a woman's high-heel shoe would leave a print none of us would have trouble recognizing. When you find a footprint that you are in doubt about often one of your local shoe stores can help you.

It was mentioned earlier in the chapter about coating the boxes with chemicals, but lets not forget fingerprints. Whenever possible it is a good idea to dust the box for fingerprints. Here again we need the cooperation of the fire department, if their men handle the box before we dust it then we will be wasting our time. But here we must also cooperate with them and dust the box as soon as possible, so they can wind the box and put it back in service. Due to local laws which, in many cases, do not allow the taking of fingerprints of juveniles, if a juvenile has pulled the box, the fingerprints may not be useful. But a good investigator should always dust for fingerprints and should also know how to improvise: If he does find a juvenile suspect, he may offer him a drink of water and give it to him in a clean glass; then when he has finishing drinking, the glass can be dusted for prints. These prints lifted from the glass can then be compared with those lifted from the fire alarm box. This way we abide by the law and still find these prints. Use your own ideas if you wish, but remember there are many laws to protect juveniles and these must be taken into consideration during investigation. A false alarm pulled or a fire set by a juvenile is just as dangerous and costly as one done by an adult.

Keep Records

With false alarms, as with any other crime, files should be set up listing known offenders. There should not just be a file of persons convicted, but also a file of persons suspected of pulling alarms. All files should be filed twice, once in alphabetical order and again in the M. O. (method of operation) file. As in most crimes, when a person finds a successful method of committing the crime he will tend to stick to it.

Enlist Aid Of Others

While working on your case use any help you can get. If juveniles committed the crime they will often brag about it. Speak to school teachers and school-crossing guards, ask them to let you know if they hear something that can be helpful. Ask the local fire fighters to help you; they can note any cars leaving the area as they are arriving on the scene and any persons seen regularly at false alarms. In many areas, when a box is pulled, there are many engines responding from every direction and, in this case, one of them is almost sure to be passing the "get-away car," if one is used.

NEW TYPE OF ALARM BOX

A new type of alarm box making its appearance in our area is a telephone enclosed in a box mounted on a pole at various corners in the community. This telephone is a direct line to fire department headquarters and has many advantages over the old box in that the person sounding the alarm can have direct conversation with headquarters and tell exactly where the fire is and what is on fire. This gives the fire bureau some idea what equipment to send. Police may also be summoned by this phone by advising the fire department dispatcher that the caller requests the police. Opening the door to this alarm box reveals the telephone and as soon as the receiver is picked up contact is made and even if nothing is said it is considered an alarm and fire equipment is sent to the scene. The reasoning of the fire department is sound in sending equipment even if there is no conversation with the caller, as he or she may be too excited or in too much of a hurry to get back to the fire to stand at the alarm box and converse. Even if the police were sent to check a "silent" alarm first, the gamble is too great that there is a fire and too much time would be lost.

SUMMARY

1. All false alarms must be investigated.
2. Box alarms where the lever is pulled to turn in the alarm is one of the most common methods of turning in a false alarm.
3. Just about any sex or age of person has been known to turn in a false alarm.
4. Question all people in the area of the box.
5. Consider the day and time the box was pulled in an attempt to eliminate suspects.
6. Coat the handles of boxes in a trouble area with invisible dye.
7. The telephone is another method used for turning in false alarms.
8. Question all people in the area where the fire was reported.
9. Question the fireman that received the phone call for information such as sex of the caller, speech defects, and background noises.
10. If the phone call was recorded on a tape recorder, review this tape.
11. Check area of box or suspected phone for physical evidence such as footprints or fingerprints.
12. Keep and review records of known offenders.
13. Enlist the aid of fire fighters and anyone else who may be able to help you.
14. As of 1964 there is a new type of alarm system which is a combination box alarm and telephone which is making its appearance in various areas.

Chapter Fourteen

POST MORTEM EXAMINATION
POST-MORTEM CHANGES

CAREFUL post-mortem examination of a fire victim is the nucleus of arson investigation when human life is destroyed. An alert investigating officer and a well-trained medical examiner can usually obtain, through examination of the cadaver, almost enough evidence to reconstruct the fire scene as to its cause, time and extenuating circumstances which either directly or indirectly caused death. What appeared at first to be an accidental-fire victim may later be proven the victim of homicide or self inflicted injury. Similarly a suspicious criminal act may be substantiated as accidental death following a careful post-mortem examination.

General Considerations

Death is followed by some early changes in the body with which it is necessary to be familiar. By definition, death occurs when both respiration and heart action cease. Individual cells or tissues may remain "alive" for variable but short periods of time after death, but very soon, irreversible changes occur such as cooling of the body, development of muscular rigidity, gravitation of blood to dependent parts, clotting of blood and putrefaction. Although varying to some extent with external and internal conditions, these post-mortem changes develop with time relationships which may be useful in roughly estimating the time of death for medico-legal purposes.

Cooling Of the Body

Cooling of the body (algor mortis) gradually occurs to the temperature of the environment, usually being complete in about

[161]

forty hours. The rate of cooling varies with factors such as environmental temperature, clothing of the body and state of the individuals nutrition at the time of death. It tends to be about 3 to 3.5 degrees F. per hour for the first few hours, gradually decreasing to 1 degree F. per hour until environmental temperature is reached.

Muscular Rigidity

Muscular rigidity (rigor mortis) develops soon after death due to a chemical change in muscles of the body. It begins first in involuntary muscles such as those in the smaller structures of the body and then is noticeable in the voluntary or large muscles of the head and neck, and gradually spreads over the entire body. It usually begins in about four to ten hours and passes off in three or four days. The time of appearance and the degree of rigor mortis are affected by a number of conditions so that it is not very reliable as an indication of the exact time of death. Violent exercise (or struggle), exhaustion, and high environmental temperature before death tend to hasten rigor mortis, whereas low temperature and a frail sickly condition tends to retard its development. Therefore determining the degree of rigor mortis in a body where death was related to fire may prove of little value in itself. Nonetheless, when considered in its proper prospective with environmental factors, rigor mortis is a valuable adjunct in estimating time of death.

Post-Mortem Staining

Post-mortem staining (livor mortis) (post mortem lividity) is due largely to changes in the position and condition of the blood. An irregular reddish or bluish discoloration of dependent parts of the body develops, due to the gravitational settling of blood. Internal organs, such as the lungs, are effected as well as the skin. Up to ten or twelve hours after death, the position of the blood, an hence the site of livor mortis, will change with variation in the position of the body. This emphasizes the importance of noting the position of the body when found, since finding the presence of livor mortis in an area contrary to the dependent parts of the body when first seen at the scene could indicate that the body was relocated a considerable time after death had occurred. This could possibly be the first indication of "foul play," and stimulate the search for further evidence of criminal act.

Clotting Of Blood

Clotting of blood occurs early after death and, in cases of slow death, may actually begin before respiration and circulation have ceased. The post-mortem clot or clotting that occurs after death is a red, elastic or jellylike clot which does not adhere to the lining of the blood vessels. Clots which begin before death are layered due to the fact that there has been time for gravity to separate the elements of the blood before clotting is complete. This is the basis on which death due to a sudden coronary (heart attack) is determined. From this layering process, the time of formation of the clot can be approximated.

Putrefaction

Putrefaction in the death body follows entrance of decay-producing bacteria into the body tissues, usually from the intestinal tract or bowels. It results in the production of gases (e.g., hydrogen sulfide) and a greenish discoloration of the tissues from the reaction of these gases with the iron of the blood. This discoloration is seen first over the abdomen and gradually spreads to more distant areas as the body decays. The hydrogen sulfide gas given off by this process of decay is what produces the sickening odor of death.

SIGNIFICANCE OF TRACE EVIDENCE

Investigations of injury or death under suspicious or unusual circumstances are frequently complicated through thoughtless acts which destroy evidence prior to the arrival of trained investigators. A police officer is often one of the first persons to have contact with the victim or assailant or agent responsible for arson. If he is aware of the significance of trace evidence and preserves it as much as possible under the circumstances, he can greatly facilitate the investigation and possibly produce clues which ultimately lead to the exact causal relationship of the fire to crime, suicide, negligence. If he is unaware of the implications of these traces, he may inadvertently destroy important evidence.

It is common knowledge that a gun or other weapon can be an essential item of evidence in an investigation of injury or death by violence. However, few people have more than a vague concept of the elements contributing to the significance of a weapon as evidence. Unfortunately the ignorance does not prevent curious persons and unauthorized amateur sleuths from handling such evidence, since they are oblivious to the consequences of this careless act. In many investigations the relative positions of various items at a scene is a factor of extreme importance. It is highly improbable that the person who carelessly picks up a suspected weapon would carefully note its position. Therefore, by merely displacing it he will distort a fundamental element of the weapon's significance.

Evidence that can be detected by unaided physical senses is appropriately described as *physical evidence*. However, the significant evidence is very often a small amount of material *(trace evidence)* that may not be apparent to the untrained observer. These traces might represent vestiges of someone or something once present and afford clues by which it is possible to trace the movements or identity of persons or objects associated with these materials. Trace evidence, then, is any material that may be used as direct or corroborative evidence to aid in establishing the facts. Such evidence may contribute significantly to proof, but it itself is not proof. The clues obtained by examination of trace evidence must be corroborated with other information and interpreted accordingly.

Trace evidence in the form of blood, body tissue, hairs, clothing fibers and prints on instruments used in crime is equally as significant as the weapon itself. These materials aid in demonstrating association of the weapon with the injured person. Careless handling of the weapon is likely to dislodge this type of evidence and render it useless as significant evidence. Obviously, the most important factor for preserving trace evidence is realizing ahead of time that anything could be a potentially valuable clue and taking precautions not to overlook such possibilities.

The variety of trace evidence is infinite and unpredictable. It cannot be stated categorically that certain types of evidence invariably will be present or always be significant in any specific incident. Blood stains may be absent or relatively unimportant in the investigation of a violent death but may be present and highly significant in the investigation of a fire or burglary.

PRECAUTIONS IN PRESERVING EVIDENCE

Firemen, policemen and other persons who would be likely to be first on the scene of a possible crime could facilitate the investigation of injury or death if they were fully aware of the significance of marks and signs of violence and associated evidence on or in the clothing or body. Obviously, care of the patient rightfully has priority, and ministrations to his welfare should not be delayed. However, routine precautions for preservation of evidence can expedite the official investigation without hindering treatment. Several examples of such precautions are worth mentioning.

Remove clothing as carefully as possible. If it is necessary to cut clothing to remove it, such as in a fire, avoid cutting through rips, lacerations or holes. Such defects, if produced by a bullet or weapon can provide significant clues. For example, distribution of gunpowder residue around a bullet hole affords evidence of the distance and angle of fire; if the hole is distorted, evidence is also destroyed.

Buttons should not be removed from garments. If done so purposely or accidentally in hastly removing clothing for purposes of rendering medical aid, they should be kept with the garment. Such items recovered at a scene of injury can be compared with similar items from clothing of other persons to demonstrate association.

Foreign matter such as bullets and fragments of metal, glass or wood removed from clothing or found nearby should be carefully preserved and delivered to the proper authorities.

Garments must be protected from contact with clothing

of other persons. Clothing of suspect and victim must be kept separate from each other at all times.

Trousers should be folded carefully to prevent loss of dust, paint chips or other material in the cuffs. Turning the pockets inside out should be prohibited. Such evidence could later be used to determine the place of death in a case where the body was relocated or placed in a fire to destroy evidence of homicide.

Stained garments should be placed on hangers as soon as possible and not rolled into a tight bundle. This will prevent obscuring of patterned imprints and dislodging of crusted stains or similar particles.

I hope these examples illustrate the meticulous care required for the recognition and preservation of trace evidence on clothing. Law enforcement agents, physicians, and other investigating officials must cooperate in this endeavor.

Significant evidence on the body of the injured is of obvious importance and must be preserved or recorded when first noted. Inspection of the hands of a victim often reveals important clues. If a wound was self-inflicted by a hand gun, blood spatters and fragments of tissue may be visible on the hands. Tests to detect gunpowder residue can be performed. Bruised knuckles, scratches, and wounds that seem insignificant may be valuable clues to defensive or offensive acts. Obviously, material adhering to or clutched in the hands must be carefully preserved. Perhaps less obvious is the importance of examining the fingernails and carefully removing any debris from underneath the free ends of the nails. The importance of noticing stains on the hands is exemplified in the following case.

A man reported to police that he shot and killed his wife in self-defense. The police found the victim lying on the floor of the bathroom with a pistol in her right hand. Orange-red stains were noted on the ends of the right thumb and index fingers, on and around a slight abrasion on the back of the left hand, and in the wash basin. The pistol in the victim's hand was not loaded and had not been fired recently. Evaluation of all evidence, including the course of the bullet wound and stains on the hands, directed the conclusion that the victim was applying Mercurochrome® to the left hand when the husband shot her. The pistol found in her hand was a war souvenir that had been placed there in an attempt to substantiate his plea of self-defense.

It is not to be expected that emergency treatment of a fire victim could be delayed to permit inspection of hands through a magnifying glass. However, protection of the hands will until such inspection could be conducted primarily because it is not thought of in the first place. Keep in mind that investigation into the nature and circumstances of injury or death in fire, or any other crime for that matter, requires observation of many details that seem trivial to the firemen concerned with controlling a burning building and preventing further bodily injury or property damage. However, being the first persons to see the victim, they are charged with an important share of responsibility for the progress of the investigation. The same is true of the physician, policeman or other individual who may be first on the scene. He must train himself to be observant and automatically observe such things that are discussed here as *trace evidence*. It is not expected that he should add to his obligations the responsibility to conduct investigations over and above his professional training. This is the specialized domain of the trained investigator and medical examiner. However, in order not to hinder such investigations, it is essential for him to recognize the need to protect evidence. When it is necessary to alter such evidence, he has the responsibility to record this alteration accurately and describe the original condition in the recording.

PHYSICAL AGENTS IN THE CAUSATION OF DEATH

Bullet Wounds

Injuries produced by gunfire have certain peculiar and important characteristics by reason of the velocity of the wounding missile. The extent of such an injury is characteristically much greater than would be expected from the diameter of the bullet. The force of the impulse is projected radially from the path of the bullet with such intensity that it may and usually does cause a cylindrical zone of destruction in the tissues that surround the main tract of the wound. Thus, blood vessels may be torn or bones may be broken at a considerable distance from the path of the bullet. One peculiar feature of injuries by gunfire is the creation of one of several secondary wounding missiles in the event the bullet strikes a bone. The fragments of the shattered bone tend to be propelled in a cone-shaped course in the direction of the flight of the bullet. Each fragment then becomes a separate destructive missile itself.

If the muzzle of a gun is in contact with or very close to the skin at the moment of fire, the rapidly expanding gases may enter the tissues with explosive effects. The structure of the tissue is such that the explosive force can be decompressed internally, such as with wounds of the chest or abdomen. The skin wound may be small. If, however, the tissue is compact and the shot is fired at contact range (as to the head), the explosive force cannot be decompressed internally because of bone barrier and the skin will usually be extensively lacerated in a stellate (star-shaped) fashion.

When a bullet leaves the body it is traveling at a slower speed and is usually tumbling; instead of producing a pushed-out exit wound it produces an irregular lacerated wound, the edges of which are everted and and the dimensions of which are usually considerably larger than those of the entrance wound.

In the case of bullet wounds produced at close range, but not at direct contact, various other alterations of the skin in the region of the entrance wound may result from the muzzle blast of smoke and flame. Recognition of burning or powder deposits may be of great medicolegal importance in relation to the range of fire as is recovery of the bullet, fragments of the bullet, or empty cartridge casings.

Death Caused By Changes in Atmospheric Pressure

The human body tolerates an increase in atmospheric pressure better than it does a decrease of equal magnitude. The rate at which a decrease in atmospheric pressure takes place is extremely important unless atmospheric pressure is lowered slowly, whether it be from a high to a normal level or from a normal to a low level, bubbles of nitrogen form in the blood. These gas bubbles thus released occlude or block small blood vessels and produce the syndrome of areo-embolism, also called *the bends, caisson disease, the staggers or the chokes*. Death from aero-embolism may occur within minutes or hours after the onset of decompression.

Blast Injury

This term designates the destructive effects of the sudden changes in pressure that result from an explosion. This is particularly interesting in cases of a victim found dead in a fire, in which case an explosion from a faulty furnace or gas leak could not only start the fire, but also caused the death of the victim.

In the case of an air blast (explosion) the compression tends to be unilateral (principal effect is on the side of the body that faces the explosion). Multiple lacerations of the lungs are commonly sustained with intra-alveolar bleeding (bleeding into the small air-sacs of the lungs); a prominent feature of the post-mortem findings.

Diffuse injuries of both the chest and abdominal organs may occur and may be sustained *with little or no external evidence* of trauma.

Death By Heat and Cold

Despite man's ability to survive wide variations in environmental temperature, his internal temperature must be maintained within a narrow range. Thus, cellular injury or death occurs if tissue temperature is maintained at a level more than 5 degress C. above or more than 15 degrees C. below that which is normal for the blood.

Exposure To Heat Or Fire

When a body is found in a fire scene, one of the prime investigative determinations includes whether the victim was dead prior to the onset of the fire or whether his death was subsequent to the fire. This can usually be determined without difficulty by microscopic examination of the internal organs and tissues, since exposure to heat while alive causes characteristic changes in certain tissues which stops immediately after death occurs. Probably most characteristic are the effects of smoke and flame inhalation upon the trachea (windpipe) which, of course, would not occur if death preceded the fire.

Exposure To Cold (Freezing)

If the area of the body exposed to cold is relatively small, a severe local injury may be sustained (frost bite) without significant lowering of the blood temperature. If the area of exposure is large, the body temperature may be lowered sufficiently to cause death from circulatory failure even though no local injury has been sustained. There are no anatomic changes that can be regarded as characteristic of death due to freezing. Post-mortem examination of persons who have died from exposure to cold may disclose nothing more than a moderate degree of heart enlargement and lung congestion.

Death By Electricity

Injury due to electricity requires that some part of the body be interposed between two conductors in such a manner as to complete an electrical circuit. Thus, a live wire may be touched with impunity so long as such a contact does not complete a circuit and result thereby in the flow of electric current through the tissues.

The path of a current through the body tends to follow the most direct route between the site of entrance and exit. In flowing through the tissues an electric current may cause death by one or a combination of several effects:

1. Cells may be destroyed directly by heat or electrolysis.
2. The current may stimulate strong muscular contractions and thus inhibit vital body functions such as respiration and heart contraction.
3. The current may inhibit the function of any vital centers or organs that lie in the path of its flow by electrical interference.

ELECTRICAL BURNS (ELECTROTHERMAL INJURY): Most of the resistance offered by the human body to an electrical current is that of the skin and the amount of surface contact between skin and external conductor. Thus, electrothermal injuries are ordinarily limited to the skin and the immediately underlying tissue.

If the contact area of the skin and an external conductor is large, the generation of heat may be too low to produce a burn and yet the amperage may be more than enough to paralyze respiration and cause death. On the other hand, if the skin contact is small, such as may occur by touching the end of a live wire, the amount of heat generated in a few millimeters of skin may be sufficient to produce a burn even though the total amperage has been insufficient to cause a significant degree of electrical penetration.

Electrothermal burns have a consistent gross or microscopic characteristics by which they may be distinguished from other burns. Arcing of the current may produce pitlike defects on the surface of the skin that are rarely, if ever, produced by other forms of heat. Metallic constituents of the external conductor may be deposited in or on the surface of an electrical burn and their presence may help to establish the kind of an electrode with which the skin was in contact.

Apart from injuries due to heat production and the explosive effects of currents of extremely high voltage there are no tissue changes that can be regarded as characteristic of electricity. Attention has already been directed to the fact that an electrical current flowing through the brain may cause death by inhibiting the areas of the brain that control respiration and that electricity flowing through the heart may cause death by stopping the heart and thus completely stopping blood circulation. In neither circumstance does the passage of the current result in characteristic alteration of the internal tissues that could be determined in the autopsy.

Death By Poisons

A discussion of poisons and poisoning is in itself a lengthy subject and is mentioned here only for completeness and to alert the reader of the necessity in considering it as a cause of accidental death or homicide.

About 10,000 deaths from chemical injury take place annually in the United States. Although an extremely wide variety of substances may cause chemical injury, statistics indicate that most cases of fatal poisonings in man are caused by a relatively few compounds. Among the more common of these are ethyl alcohol, carbon monoxide, anesthetic agents, barbituates, and heavy metals such as arsenic and lead.

Approximately 55 per cent of all deaths due to chemical agents are accidental; 44 per cent are suicidal, and less than 1 per cent from homicide.

Post-mortem examination can almost always determine that death was due to poisoning because of the characteristic tissue effects produced by the various chemical agents. Needless to say that an alert investigating officer who is suspicious of death by poisoning and advises the medical examiner of this suspicion, may greatly facilitate the post-mortem examination.

SUMMARY

1. In estimating the time of death of a victim consider the external factors such as room temperature, clothing, etc., in relation to body temperature.
2. Rigor mortis begins in the neck muscles in from four to ten hours after death.
3. The study of post mortem lividity which is the gravitational settling of the blood in a victim can be the first indication of foul play. This can indicate if a body was moved after death.
4. Preserving evidence is secondary only to your duty to the victim if signs of life still exists. If no signs of life exists do not move anything until it is properly identified and its location noted.
5. The victim's clothing is an important clue. Knife or bullet holes, powder burns, or tears from a struggle are all important clues and should be noted.
6. Whether a fire victim is still alive or not, protect the hands until a complete examination of them can be made. Evidence of what the victim was doing just before he died or was injured can be found on the hands or under the fingernails.
7. The medical examination of the windpipe is the most common way to determine if a fire victim was alive during a fire. If he wasn't alive, he could not have inhaled smoke or flame.
8. Death from electrical contact or electrical burns will be hard to define and an investigator will have to use caution to protect himself if electricity is involved.
9. In all cases where an autopsy is to be performed on a victim found in a fire, the investigator should give the examiner all pertinent information to facilitate the examination.

Chapter Fifteen

VIOLATIONS AND LAWS

EVERY state has its burning and arson laws and the investigator should be completely familiar with them. Generally speaking, these laws cover only deliberate burning—for the various reasons previously mentioned. Insurance, grudge fires and fires to cover up another crime, as well as those set by pyromaniacs and juveniles are all covered in the arson and burning laws. This does not mean that fires due to other causes should all go without prosecution.

Controlled Burning

If a destructive fire was found to have been caused by burning trash, the local or state laws governing this type of fire could have been violated. These laws vary in different areas and at different times of the year and depending on the location. A person should definitely be prosecuted if an emergency ban was put on burning due to drought conditions and violation was responsible for a fire. In areas that permit controlled burning, it is required that responsible persons be present during the burning until the last spark is out. Also, a fire break is required to prevent the fire from spreading. This will consist of a cleared area or some method to contain the fire. If a person wishes to destroy an old building it is required in most areas that permission be obtained from the fire department. As a matter of fact the making of any large fire in most areas requires that the fire department be notified.

Steam Locomotives

In some areas railroad steam locomotives must be provided with appliances to prevent the escape of fire and sparks from the smoke stacks when operated through forest or brush areas. It may also be required that a safety strip be cleared on both sides of the track through all forest and brush areas. This is an area of a given width that is cleared of all flammable materials so that if the train does give off sparks or fire there will be nothing along the track that can burn.

Utility Companies

Ordinances and laws established in conjunction with the utility companies can be the cause of a fire when violated. A person using a jumper or bypass on a gas or electric meter when the service is turned off, or bypassing the meter to keep it from registering, creates a dangerous condition and will usually be prosecuted by the power company when exposed. When an electric meter is bypassed the fuses are sometimes included in this bypass thereby eliminating the protection offered by the fuses. When electrical trouble is involved in a fire, always examine the meter and the fuse boxes. Bypassing a gas meter involves a little work and a person doing this will use a piece of old hose to prevent having to go through the trouble of cutting and threading pipe. In most areas, the use of hose on gas lines is in itself a violation. When any violation of this type is uncovered, it is best to notify the utility company involved.

Storage of Flammables and Explosives

The storage of flammables and explosives are, in most localities, highly controlled even to the type of container used and the temperature of the room in which they are stored. Other items that are susceptable to spontaneous combustion, such as peat moss and hay, are also rigidly controlled. Peat moss should be stored outside; even half a bucket stored in the cellar of a home has been known to begin to smolder from spontaneous combustion and fill the house with smoke. In this particular case there was no damage other than smoke, but it did cause the activation of the fire department.

FIGURE 45. This photograph shows the rear view of a variety store. The front and sides were of brick and masonry construction, but the rear was mostly wood and shingle. The lowest level shown in the picture is below the ground level in the front. This area was used for storage. The windows on the bottom level in the rear were boarded up due to a false wall inside the building. Stored in this building was a large variety of items such as plush toys, drugs, bicycles, wagons, household items, many types of plastic items, and many assorted novelties. The middle level shown was the actual store; the top level was an office-living type area, with some storage. This fire started in the middle of the night and completely gutted the building.

Business and Public Building Violations

Industrial establishments which store oily rags, excelsior and other highly flammable materials are required to use metal containers. Even improperly stored trash can be in violation because it is a potential fire hazard. Also, the blocking of or inadequate fire exits should be checked by investigators, especially if there has been injury or loss of human life. The condition of fire exits, stairways, and hallways, should be mentioned in your report. Even if there is no prosecution, there could be a civil suit

FIGURE 46. This is the ceiling inside a kitchen of a business place. A very hot fire had taken place below this ceiling. Other than the paint peeling and burning, very little damage was done to the ceiling. This was a fire-resistant ceiling placed in the kitchen in compliance with the fire prevention codes in the area. If it were not for this ceiling, the fire would undoubtedly have spread to the whole building. This points out the need for the fire investigator to have a working knowledge of the construction of buildings and the qualities of different materials. This fire started in the middle of the night when the business was closed and the owner was asleep over it. There is a great possibility that this ceiling saved not only the business but the owners life. The good construction of the kitchen along with good work by the fire department confined the fire to the kitchen.

later on in which you may be called upon to testify about a condition that you cannot remember unless you record it. Violations of the building and electrical codes are common in businesses as well as in homes. These departments rarely have enough men to police all the building and improvements in progress. Persons who need another room or some additional wiring will often take it upon themselves to do it, without consulting the proper departments and learning the minimum requirements and type of material the law requires. In electrical installations there is, in most cases, a minimum gauge wire to use as well as a particular type armor for the wire or cable. There are also specifications for the outlet boxes and the number of outlets on a circuit. In relation to fires, building codes are just as important as electrical regulations; they control the material used and the distance it must be from furnaces, heat pipes and chimneys. Kitchens in restaurants and doors in hotels all have certain specifications for fire resistance. In most areas, even curtains in movie theaters are controlled by laws. Schools, factories and other multi-floored public buildings have fire department regulations governing which hall or stairway doors must be kept closed to prevent a flash fire. Fire in the basement, given the proper draft, can spread to all floors in minutes. When investigating a fire in a large public building damaged on more than one floor, take into account the draft condition. If the fire department response was immediate and the fire had no chance to burn through the floors, it is probable that someone left a door or two open which should have been closed. Almost all public buildings or businesses which are open to the public come under special regulations for the protection of the public. Restaurants, dance halls, bingo halls and other places of public gathering will have capacity regulations as well as a minimum number of exits. Also, all exit doors in public places should open out and may be required to be equipped with a panic bar. The panic bar is a bar fitted to the door on the inside so the door will open out from the pressure of a panicky crowd pushing against it. Exit lights are also required in public buildings in many areas.

In movie houses there have long been established laws concerning the projection booth and the storage and handling of the

FIGURE 47. This is the front of a movie theater which burned. This type of building has many special problems for the firefighter and the investigator. The projection booth with its film and other equipment and the large open area in the theater itself cause special problems. This building was of concrete block construction, but you can see that the wood around the doors, windows and ticket booth burned rather well. Fortunately, this fire started in the middle of the night when the movie was empty. The movie had the proper exits but panic is almost always a problem with crowds of people.

highly inflammable movie film. Also the balcony should have adequate egress including a fire escape capable of handling a minimum number of persons at one time. This usually means a complete fireproof stairway built up to the balcony outside of the theater which is wide enough to accommodate a number of persons comensurate with the capacity of the balcony. In many areas, sprinkler systems are required to be installed and/or adequate fire extinguishers or hoses. We all know the value of an automatic sprinkler system. This is a system that actuates automatically due to an excessive temperature increase. In many cases this type of system will have the fire contained and under control upon arrival of the fire department. Having all this required equipment installed is not the end of complying with the law—there is still the question of maintenance. Empty fire extinguishers and corroded water-valves are of no use in case of a fire. Or is a sprinkler system that has been carelessly painted over when the ceiling was painted. Careless painting can also seal fire doors and obliterate the instructions on fire extinguishers.

Transporting Flammables

There are similar regulations for transporting flammables which apply to all types of carriers. Tunnels and many congested areas prohibit some types of explosives and flammables to be transported through. In most areas motor vehicles carrying this type of cargo are required to have the vehicle plainly marked on the outside. This is a warning for other vehicles to give the transporter extra consideration and in case of a mishap, the emergency crews on the scene know with what they have to contend.

Model Arson Law

Each investigator must study the arson laws of his state. The *Model Arson Law* which is described here is in use in most states.

ARSON— (first degree) BURNING OF DWELLINGS

Any person who willfully and maliciously sets fire to or burns or causes to be burned or who aids, counsels or procures the burning of any dwelling house, whether occupied, unoccupied or vacant, or any kitchen, shop, barn, stable or other outhouse that is parcel thereof, or belonging to or adjoining thereto, whether the property of himself or of another, shall be guilty of arson in the first degree, and upon conviction thereof, be sentenced to the penitentiary for not less than two nor more than twenty years.

ARSON— (second degree) BURNING OF BUILDINGS, ETC., OTHER THAN DWELLINGS

Any person who willfully and maliciously sets fire to or burns or causes to be burned, or who aids, counsels or procures the burning of any building or structure of whatsoever class or character, whether the property of himself or of another, not included or described in the preceding section, shall be guilty of arson in the second degree, and upon conviction thereof, be sentenced to the penitentiary for not less than one nor more than ten years.

ARSON—(third degree) BURNING OF OTHER PROPERTY

Any person who willfully and maliciously sets fire to or burns or causes to be burned or who aids, counsels or procures the burning of any personal property of whatsoever class or character (such property being of the value of twenty-five dollars and the property of another) shall be guilty of arson in the third degree and upon conviction thereof, be sentenced to the penitentiary for not less than one nor more than three years.

ARSON— (fourth degree) ATTEMPT TO BURN BUILDINGS OR PROPERTY

(a) Any person who willfully and maliciously attempts to set fire to or attempts to burn or to aid, counsel or procure the burning of any of the buildings or property mentioned in the foregoing sections, or who commits any act preliminary thereto, or in furtherance thereof, shall be guilty of arson in the fourth degree and upon conviction thereof be sentenced to the penitentiary for not less than one nor more than two years or fined not to exceed one thousand dollars

(b) Definition of an attempt to burn

The placing or distributing of any flammable, explosive or combustible material substances, or any device in any building or property mentioned in the foregoing sections in an arrangement or preparation with intent to eventually willfully and maliciously set fire to or burn same, or to procure the setting fire to or burning of same shall, for the purposes of this act constitute an attempt to burn such building or property.

Burning to defraud insurer

Any person who willfully and with intent to injure or defraud the insurer sets fire to or burns or attempts so to do or who causes to be burned or who aids, counsels or procures the burning of any building, structure or personal property, of whatsoever class or character, whether the property of himself or of another, which shall at the time be insured by any person, company or corporation against loss or damage by fire, shall be guilty of a felony and upon conviction thereof, be sentenced to the penitentiary for not less than one nor more than five years.

We have covered the Model Arson Law because it is used in most states at this time. It is covered here for use as general information and it is very important that the investigator study the laws of his own area.

SUMMARY

1. The investigator must know the arson and burning laws of his state.
2. Most areas have laws covering controlled burning of such things as trash, leaves, old autos, etc.
3. There are special laws that cover the tampering with the meters and other equipment of the gas and electric companies. The bypassing of a gas or electric meter can often be the cause of a fire.
4. In most areas, there are many restrictions on the storage of flammable and explosive items.
5. There are many building and electrical codes that will be of interest to the fire investigator.
6. Public buildings will have many restrictions as to the number of people permitted in the building, fire escapes, exits, and fire extinguishers.
7. There are regulations covering the transportaiton of flammables and explosives.
8. The Model Arson Law, which covers willful and malicious burning, is used in most states.

Chapter Sixteen

SCIENTIFIC INVESTIGATION

THERE are many scientific aids to fire investigation, some of which can be applied at the scene and others which can be better carried out in a crime laboratory. Most investigation organizations have a crime laboratory at their disposal and investigators should know what services are available and how to apply them. Lifting, studying and identifying fingerprints and footprints as well as microscopic examination of matches, broken glass for threads of clothing and chemical analysis to determine the presence of an accelerant are a few of the scientific processes.

In the case of arson, let us first elaborate on the information that can be obtained from footprints found at the scene. If the arsonist gained entry by breaking in through a window, the footprints outside the window will most likely be his. This would not be the case if he enters by forcing a door that was in frequent use. That the window was broken from the outside can be confirmed by examining the edge of the glass. There is no set rule for a burglar or arsonist to leave the building the same way he entered. Our experience has shown that he will often leave by a rear door and sometimes by the front door. In this case, you will find many other footprints and you will have to consider those found at the point of entry and look for a similar set leaving the scene. The footprints made by the arsonist leaving the scene should be further apart and may be deeper at the toe. He may not be running because that would arouse suspicion, but he will not be wasting any time as he did when he crept up to the building because his job is finished. Also, if the arsonist used an accelerant it is possible he may have spilled some on his shoes or walked in it. This will penetrate his shoes and leave tracks. This is important to remember when picking up suspects.

Footprints and Fingerprints

There are many commercial products on the market for making casts of footprints. This should be done slowly so as not to disturb any of the detail of the footprint and the cast should be reenforced to keep it from breaking. Footprints on a hard surface will have to be photographed and should be done at an angle. As a matter of fact even indented footprints can be photographed before they are cast. Details of a footprint may show up on a photograph which are hard to see on the cast. The FBI keeps shoe print files which can help determine the trade name of heels and soles. Unfortunately, the authors have found footprints easier to find than fingerprints which would be a more positive identification of an arsonist. This is because the heat and smoke (or the fire hoses) obliterate any fingerprints. However, fingerprints should not be disregarded entirely, as a piece of glass thrown outside by the arsonist, at the point of entry, or any objects moved by him (especially outside the building) may reveal good fingerprints. Inside the building, smooth objects in closets or drawers that were handled by the arsonist and then replaced may contain good fingerprints which were protected from the fire. If the investigator arrives on the scene of a suspected arson in the early stages of the fire, he should remove any small objects that may have fingerprints and record their location before the fire becomes too intense. Most arson investigators are not fingerprint experts and the lifting of these prints should be left to the experienced. Of course, if a fingerprint is in danger of being obliterated and cannot be removed for safe keeping a photo of it can be made then an attempt to lift it by the investigator.

Photography

The constant mention of photography in this chapter is indicative of its relation to scientific investigation. Not only are photos taken at the scene of a fire important but those taken in the laboratory of results of test comparisons of materials. All of these can be used in court. Many microscopes have camera attachments which are indispensable for recording evidence seen only through a microscope. It is the opinion of the authors that

FIGURE 48. The lumber yard fire is a very serious and dangerous one. In the rear of this picture you see a pile of lumber that is completely engulfed in flames. The wind is blowing to the left of the picture. The open area between the pile which is burning and the piles of lumber shown in the foreground of the picture, along with a good job by the firefighters, kept this fire from spreading. Sparks are a great problem in this type of fire due to the large amount of sawdust that is usually on the ground. If you have more than one fire at the same time, this must be kept in mind: Consider the size of the fires, distance between the fires and the wind direction.

when an investigator arrives at the scene of a fire, regardless of its origin, he should have his camera in his hand ready to use; the scene he sees at that time can never be duplicated. Among the important facts that will be recorded by the camera, if early and frequent pictures are taken by the investigator, are the general location of the origin of the fire, the wind direction, which has considerable bearing on the spread of the fire. A few shots can also be made to include spectators for future reference. The times that the photos were taken will have to be recorded also. Due to all the work involved in making a complete and efficient fire investigation, it is best that two men work together and coordinate their activities. One man should carry a small high-quality camera with flash attachment which he can carry in his pocket. The problem with this type of camera is in taking pictures at night the small flash is sometimes inadequate to light as large an area as you would like to photograph. This can be overcome by borrowing the bright portable lights that the fire department will have at the scene. Of course, if your organization has an unlimited budget, a professional photographer using the best color equipment would be an asset to your investigation. Be sure to keep your flash equipment in working order and carry spare batteries. If possible have one investigator test the flash equipment while enroute to the fire. Many good shots have been lost because flash equipment failed to function and by the time it was made operative it was too late. Also carry ample spare flashbulbs and film.

Improvising

There is a limited amount of equipment that an investigator can carry on his person while making his investigation so he will have to know how to improvise. Small items like tools, match folders, cigarette packs, etc., that are suspected of having been left at the scene by an arsonist can be picked up by the investigator and after recording their location may be placed in any container to protect them. If an immovable object has fingerprints on it and is in danger from the fire, the investigator can bring out the prints by sprinkling any dust, powder, or ashes

FIGURE 49. This picture shows an example of evidence of a forced entry at the scene of a fire. This is a basement door, located in the rear of a home. The fire was located on the first floor and the fireman entered the home by the first-floor doors. The molding on the outside of the door was pried off, so the arsonist could reach door latch. A close look at the picture will reveal the pry marks on the molding. In this case, they were of the type that would be made by a screwdriver. If you look closely you will also see that the pane of glass closest to the door knob is missing. This is because there was a bolt lock on the inside of the door and the arsonist had to break the glass and reach in to open it and thus complete his entry.

available on the prints and then blowing off the excess which should bring out the prints enough to photograph. The print can then be lifted using transparent tape. An investigator should practice this procedure using different substitutes for the powder and observing what substitutes bring out the prints best on different objects. Also, the process of lifting the prints quickly will require some practice.

If forced entry is made to the building by using some type of pry bar it is a good idea to remove the piece of wood that was pried and preserve it as evidence. Later you may find the tool that you think was used to make the forced entry and wish to check the tool marks against the wood. Each tool has individual characteristics and will, in most cases, leave distinctive markings which can be identified in the laboratory. Often paint, wood, glass or other substances from the crime scene may also be found on the tool when it is examined in the laboratory.

It is possible that dirt from the suspect's shoes, pants cuffs or any other part of his clothes or body can be compared with that found at the scene. Even the dirt under his fingernails should be collected and examined, as it may compare with that at the crime scene and it may even be found that there are traces of an accelerant present or perhaps wax from a candle.

If an explosive was used in the offense then it is good to know that the FBI has a reference collection of dynamite wrappers, fuses and blasting caps.

Stains

When a stain found at the scene is suspected of being blood a laboratory examination can determine if it is blood and if it is animal or human blood. If a large enough sample is supplied, the laboratory can also classify the blood.

Burned Records

In some cases, you will find that records and other papers that you would like to see have been burned in the fire. If these burned papers are properly preserved it is possible that they can be restored and read. As you can see there are many types of fires where this can prove to be very helpful.

The authors have seen cases where forced entry has been made by breaking windows and reaching through the jagged glass to unlock the window or door. In doing this the subject cut himself, leaving blood at the scene, and also left a piece of skin on the glass. This, of course, gave information as to the race of the subject wanted for the offense.

Matches

It has been pointed out that any matches found at the scene should be preserved as evidence. This is important because you may pick up a suspect that has a partly-used match folder in his possession. These matches can then be compared with the folder under a microscope in an effort to determine if the matches found at the scene were torn out of the folder found on the suspect.

Cigarettes

The paper and the tobacco of any cigarette found at the scene can be examined in the laboratory to determine the brand of the cigarette. You may also, in some cases, get a saliva sample from the cigarette which could also later be compared with the saliva of a suspect.

Any unknown substances found at the scene of a fire should be examined to determine whether it is an accelerant and if so, what type it is. You can then determine who sells this product; whether there was a reason for it being in the building, and if not, who made a recent purchase of this product.

Analyzing Ashes and Remnents

The laboratory can be very helpful in cases where you feel that a substitution of merchandise has been made in a fire. An example of this could be where a check with the manufacturer of a certain type of furniture reveals that the handles on it are made of brass, but the laboratory examination of handles found in the fire shows that they are made of steel or some other metal. In many cases an examination of the wood found may even show it is a different type of wood than that shown on the inventory.

You may even find that the cheaper article which burned was made of veneer while the good article that is being claimed was made of solid wood. A comparison examination can also be made of any screws, nails or other fastening devices found in the fire and those that normally would be used in the construction of the item claimed to be burned. Even if the same type of nails or screws were used in the cheaper item you may find that they were of a different size or quantity. In cases where clothing or cloth goods have been burned you can, of course, compare buttons or any metal items found in the fire with those on samples obtained from the manufacturer. Pieces of cloth found in the fire should be collected as evidence. The fibers in this cloth can be compared under the microscope with samples of known cloth to determine the type of cloth it is and if it is the type of cloth that was claimed to have been burned in the fire.

This type of examination is also helpful when you find fibers from the criminal's clothing at the point of entry or elsewhere in the building. This fiber can give you information as to the color of clothing the subject was wearing at the time of the offense and also the type of material his clothing was made of. When you do pick up a suspect these fibers can then be compared to the fibers of his clothes.

Identifying the Suspect

Any hairs found at the scene can also be examined to determine the color of the hair of the person you want. These hairs can also be compared under the microscope with the hairs of a suspect. We might point out that most of this type of evidence by itself is not enough to get a conviction, but it is strong evidence when used along with other evidence that you collect. And it does prove to you as you are working on the case that you are on the right track as it will eliminate the persons who are not guilty. In many cases this type of evidence is enough to get you a confession if, when questioning the suspect, you confront him with some of this evidence.

Identifying His Motor Vehicle

Very often, the fire setter will use an auto or truck in his work. Therefore, it is possible that you will find tire tracks at or around the fire scene. These should be photographed immediately and casts made as soon as possible. If the casts cannot be made right away then be sure to protect the tire tracks so they will not be destroyed. If you do not protect the tracks, you may find that when you are ready to make your casts, you have a very good set of tracks made by the tires of the local fire engine. (Don't blame the driver of the fire engine, it's your own fault!) A good set of tire tracks will often reveal the trade name and the size of the tires. Examine each track carefully and make as many casts as necessary—you may find that each wheel on the vehicle has a different make of tire on it and, in some cases, you will find that the tires vary in size.

When an auto is used in an offense it is not uncommon for the fire setter to be involved in an accident when leaving the scene. Any paint from his auto that is left at the scene should be collected and examined. From the paint sample it is possible to suggest the type of auto wanted by comparing the sample with automotive paint files. Of course, if you have a "suspect" auto then it is possible to directly compare the sample taken at the scene with a sample taken from the suspect's auto. Often the suspect's auto will pick up paint from the object it struck at the scene. The auto can then be examined for any paint the color of this object and if any is found then samples of this can be taken and examined. Any pieces or part of the suspect's auto that are left at the scene should be collected and examined. Often, much can be learned such as the make, model, year, and color of the auto wanted. If the subject struck something made of wood then often he will carry with him part of the object. When a tree is struck, the auto may take branches with it; when a house is struck, wood or shingles may be stuck in the bumper or torn fenders. As you can see any of this can be very good evidence that the suspected auto was at the scene of the fire.

SUMMARY

1. Learn to avail yourself of the services of your crime laboratory. Familiarize yourself with the records and services available through the FBI.
2. If there is evidence of another crime being involved with a fire, make every effort to protect the scene. Do not let any evidence be destroyed in the fire.
3. Practice lifting fingerprints with any available dust you may find at a fire scene (cigarette ashes, lead pencil scrapings, etc.).
4. When taking "close-up" photographs with a flash attachment of footprints or fingerprints, take an angle shot or you will get nothing but a picture of your own flash.
5. Take a look through a microscope and discover the wealth of information that can be gained from the smallest clue. Once you become aware of the possibilities of a microscope you will use it often.
6. Familiarize yourself with different accelerant odors. Gasoline, kerosene and lighter fluid are the most common accelerants.
7. Photos and plaster casts of footprints and tire prints can be valuable to an investigator.
8. Footprints, if properly examined, can give much information about the one who made them. The length of the stride, depth of the print and the angle of the feet can help describe a suspect.

Chapter Seventeen

EXPERIMENTS AND ILLUSTRATIONS

THE experiments and illustrations, which, in some cases were staged by the authors, are designed to give the investigator visual information and clues as to what to look for at a fire scene un-

FIGURE 50.

der various conditions. It will be of help to investigators to erect their own props and conduct experiments. This will familiarize them with the residue left by certain accelerants, as well as time factors in burning.

In Figure 50 the fuse box is shown. Notice the circuit diagram on the door of the fuse box. This diagram will tell the investi-

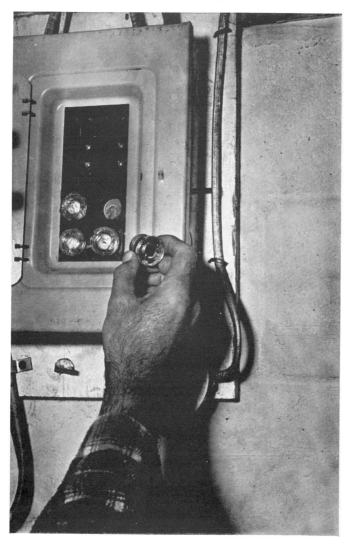

FIGURE 51.

gator where each circuit (or fuse) in the box goes. The investigator here is preparing to pull the cartridge fuses which will disconnect them from the main line and make it safe for him to remove and examine the small fuses and their sockets. Notice that the investigator in this picture uses only one hand to remove the fuse block. The best place for the other hand is in your pocket or behind your back. Never hold the fuse box or lean on the wall or other grounded object when working around electricity.

With the cartridge fuse removed, the investigator safely removes the suspected open fuse (see Fig. 51). If you look closely, you will see a penny in the fuse socket. This is a method used by people who cannot find a heavy enough fuse to keep the current flowing through an overloaded circuit. Also, being caught without a spare fuse when one is needed may cause a person to employ this dangerous device.

The diagram on the inside of the fuse-box door will tell you to what part of the house the suspected circuit leads. However, in some cases there may not be a diagram, or it may be illegible; in this event it will be necessary to completely remove the front of the fuse box to expose the wiring (see Fig. 52). With the wiring exposed, you can plainly see which wire leads from the socket with the penny in it to one of the cables leaving the top (in this case) of the fuse box. When exposing the entire inside of a fuse box, remember that the terminals at the top of the block where the cartridge fuses were is still hot. It is not easy to touch these protected prongs accidentally, but it can be done.

Directly below and to the right of the main fuse box is what may be described as a smaller fuse box. This consists of a square cover over a round fuse attached to a round plate over a square junction box. The usual purpose for this arrangement is to protect the heating plant. If there was trouble in the heating plant which is activated by electricity (gas or oil heat), look for a fuse arrangement similar to this.

The meter itself is in the glass dome on the box, to the left and below the fuse box. There are no fuses or connections in this box with which we should be concerned. However, if the power company was being defrauded by someone bypassing the

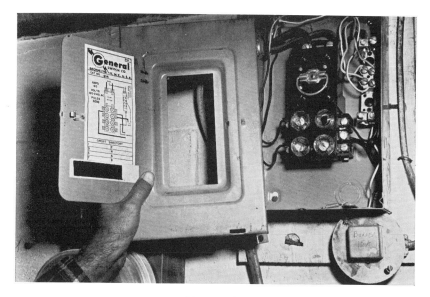

FIGURE 52.

meter, this evidence may be located around the meter and the power company should be contacted.

An investigator will have to familiarize himself with the various types of fuse boxes and meters. Some vary considerably from those pictured. Many of the older units are one piece, with the meter and the resident's fuse box together. The portion of a fuse or meter box that is not to be opened except by the power company will be sealed with a piece of wire and a small lead seal. Do not, under any conditions, break this seal. Other fuse boxes contain no cartridge fuses and are disconnected by pulling down a lever on the side of the box. Still others have the meter outside the house and the fuse box inside. Remember, when in doubt, call the respective utility company.

Timing Devices

One can readily understand that the candle and the cigarette are the most common timing devices. They are both easily obtainable and it takes no particular ingenuity to figure out their time factor. The knowledge required to make a mechanical or electric timer is not needed to use the candle or cigarette for a timing device. A simple experiment with different brands of cigarettes and different manufacturer's candles will tell the arsonist how much time, within minutes, it will take his chosen device to start a fire. Some loosely packed cigarettes will burn faster and the king size ones will naturally give more time. The speed of burning for a candle depends on the diameter and the type of wax used in its manufacture.

In our experiment (see Fig. 53) we found that the five-inch candle burned one inch every hour. The standard size 2¾ inch cigarette took fifteen minutes to burn completely while in a horizontal position.

FIGURE 53.

To further illustrate how these devices can be used, for example, a cigarette can be placed in a book of matches (see Fig. 54). In our illustration about one-third of the cigarette is left protruding beyond the match heads. This will give the arsonist about five minutes to get away. The matchbook can then be closed, or left open, and placed in some dry kindling (see Fig. 55) while the arsonist gets away.

FIGURE 54.

FIGURE 55.

Since the candle produces light and can be blown out by the wind, you can see that the cigarette is preferred for outside fires and the candle is more adaptable for indoors.

When a candle is used as a device to start a fire, some effort will be made to cover the windows so the light cannot be seen from outside unless, of course, it is put in a box or a closet. The

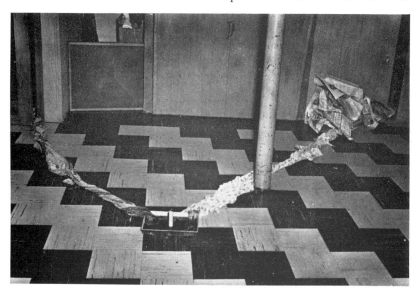

FIGURE 56.

candle will then have to be set in some kindling or a pan of accelerant (see Fig. 56). Kerosene, rather than gasoline, is preferred for this. Often excelsior or corrogated cartons are used. The arsonist may decide it is necessary to spread the fire from room to room and use oil-soaked rags (as in Fig. 56). Other materials and methods can be used to spread the fire such as papers hung across the room (see Fig. 57).

In most cases there will be evidence of the trail that can be detected by an alert investigator. It may be necessary to analyze the ashes from a trail to tell just what was used, or take up some of the flooring to find traces or an accelerant. These extreme laboratory measures, however, should not be necessary to indicate arson if a sign of a trail is clearly visible and can be photographed.

FIGURE 57.

FIGURE 58.

FIGURE 59.

Mechanical detonators are more popular for setting off bombs than fires. One reason for this is that, even after a complete fire, much of the mechanism such as gears, springs, and wheels will be left in the ashes (see Fig. 58). This is a definite indication that arson has been committed. Also, it is not practical to place a lighted cigarette or a candle in a suitcase to put on board a plane or in a locker of a building that is to be blown up. Alarm clocks, door bells, telephones, blasting caps and dynamite sticks are all tools of the arsonist or bomber who use mechanical detonators. In cases where an auto is to be destroyed, often its own ignition spark is used to set off the destructive force. This same electrical contact theory can be applied to door-bells and telephones, or any other device that will create a spark or electrical contact at a given time. In all cases where a mechanical detonator was used to set off a fire or bomb, an experienced investigator should be able to locate some familiar fragment that he can identify as a part of the timing device (see Fig. 59).

The Molotov Cocktail

This is a device which in itself may well be called a bomb with a very short fuse. It consists of a bottle or jug (see Fig. 60) containing an accelerant with rags stuffed into the accelerant and protruding out enough for a wick. It is customary to light the wick and throw the jug against a wall or into a room so it shatters. This causes a hot, fast burning fire that will spread rapidly

FIGURE 60.

and give any occupants of the building little chance to escape. The Molotov cocktail was probably made popular in wartime as a defense against armored vehicles by unarmed civilian fighters. One brave civilian fighter could run up to an enemy tank with a Molotov cocktail and throw it under the tank, which is its most vulnerable place, and completely destroy it. Unfortunately this deadly device has retained its popularity and is often used as a tool of hate groups to terrorize their opponents. While any novice investigator can suspect arson when he finds an alien accelerant container at the scene of a fire, it will take a thorough, probing investigator to locate the fragmented remains of a glass bottle or jug. Since the Molotov cocktail was meant to break, the investigator will have to probe the ruins for parts of the container. The glass particles of a bottle or jug will have a contour and thickness unlike the flat fragments of window panes and other glass in the rubble. The visual residue and odor of the accelerant will also accompany the glass particles you are looking for.

The room in which the fire appeared to have started should be considered when sifting through the debris. A broken milk bottle, for instance, is not normally found in the living room; but a broken milk bottle may well be found in the kitchen. In this case you would have to analyze the residue on the glass to see if it contained an accelerant. If enough particles of glass are found, you may be able to locate a trade name that will help to determine whether the container was alien to the location.

Particles of a one-gallon jug, which is pretty well standardized in shape and size, will be comparatively easy to locate. The neck of the jug with its heavy handle and the bottom will usually be found in separate single pieces. With either of these sections or a curved piece of a jug you can easily make comparisons with other jugs and determine exactly what size was used.

Metal Containers

While metal containers are rarely used to ignite and throw into a building to start a fire, they are the most popular type of container used to transport accelerants. The easily obtainable one-two-and five-gallon cans with screw caps are the most common (see Fig. 61). In many cases, these containers are discarded by various commercial establishments and the fire setter who uses one will more than likely leave it at the scene when he is through with it.

If the regular heavy-gauge gasoline can that has a spout and cap and costs a few dollars to buy is found at the scene of a fire, it was most likely stolen from a nearby shed or garage (see Fig. 61).

FIGURE 61.

The finding of a familiar cigarette lighter fluid can, especially where a person was burned can indicate accidental or deliberate burning. It is not uncommon for a person to fill a cigarette lighter and spill the fluid all over their hands and clothes. They will then light the lighter to see if it works. Equal danger exists for the person who uses an accelerant to help ignite a stove or an outdoor charcoal fire. If the regular container of fluid for this purpose is used however, the danger is minimized by squirting the fluid through small holes punched in the top of the can.

The familiar spray can has become a popular method of dispensing fluids. Paints, insect sprays, hair sprays and many other

FIGURE 62.

fluids are dispensed from these cans in an atomized state. Any open fire or spark is all that is needed in many cases to start a flash fire or cause disfigurement to the user (see Fig. 62). In incidents where one of these spray cans was involved in starting a fire, you can assume someone was holding it in their hand. The user would naturally drop or throw the can when the fire started and if the heat is intense enough where the can landed you will have an explosion when the airtight can is heated.

TABLE I

IGNITION TEMPERATURE AND FLASH POINT
OF CERTAIN SUBSTANCES

Flash Point: the lowest temperature at which a flammable liquid will give off sufficient vapor to support burning.

Ignition Temperature: temperature to which a material must be heated for burning to occur. There need be no spark or flame present.

Substance	Ignition Temperature	Flash Point
Acetone	1,000	0
Benzene	1,000	12
Stoddard solvent	450	100
Creosote	637	165
Ethyl alcohol	700	55
Fuel oil No. 2	494	110
Linseed oil	650	432
Sulphur	450	405

* Shown in degrees Fahrenheit.
Note: the above figures were taken from various sources.

TABLE II

HEAT PRODUCING CAPABILITIES OF SOME
COMMON TYPES OF FUEL

A British Thermal Unit (BTU) is the amount of heat necessary to raise one pound of water one degree in temperature.

Substance	BTU Per Pound
Coal, Anthracite	12,520 to 13,830
Coal, bituminous	10,020 to 14,700
Birch	7,580
Oak	7,180
Gasoline	19,800 to 20,520
Kerosene	19,710 to 19,890
Paper	6,710 to 7,830
Dynamite	2,320

Note: the above figures were taken from various sources.

TABLE III

TEMPERATURES OF INTEREST TO THE INVESTIGATOR

All temperatures shown are in degrees Fahrenheit. These temperatures are in most cases approxiate, and should be used as such.

6,500 electric arc under pressure
5,000 oxy-acetylene flame
4,500 oxy-hydrogen flame
2,900 sand melts
2,732 flames appear dazzling white
2,550 flames appear bright gray white
2,372 flames appear yellow white
2,200 cast iron melts
2,192 flames appear orange yellow
2,012 flames appear orange red
1,980 copper melts
1,832 flames appear light red
1,500 flames appear cherry red
1,400 glass softens
1,215 aluminum melts
1,141 flames appear dark red
1,000 steel loses over half of its strength
975 red flames barely visible in daylight
620 lead melts
600 wood ignites on 5 minute exposure
500 wood ignites on 10 minute exposure
425 wood ignites on 20 minute exposure
375 wood ignites on 30 minute exposure
212 water boils

Note: the above figures were taken from various sources

GLOSSARY

abet: to encourage the commission of an offense

accelerant: a substance added to speed burning

acquit: to free or clear someone of a criminal charge

addenda: things to be added

affidavit: a sworn written statement

algor mortis: cooling of the body

apparatus: any type of fire fighting equipment such as a pumper, hose wagon or ladder truck

back draft: explosion of pent up gases, caused by heat and smoke suddenly receiving fresh oxygen. Correct ventilation will usually prevent this

British Thermal Unit (BTU): the amount of heat necessary to raise one pound of water one degree in temperature

buff: a civilian who is interested in fires and fire fighting

cadaver: a corpse

cadaveric spasm: a stiffening of the arms or hands which may take place under certain conditions at the time of death (death grip)

charged line: a fire hose that has water in it

circuit: one complete line from source and return

circuit breaker: same purpose as a fuse

coercion: the act of forcing to comply to any action; forcing by use of threats, authority, or any other means

conduction: physical flow of heat through a material from the point at which the heat is produced, to remote outer portions

convection: upward physical movement of heat

coupling: fittings on fire hose used to connect one piece of hose with another

deposition: sworn testimony obtained out of court

domicile: permanent place of abode

doughnut: a rolled up length of hose

fuse: safety device to open a faulty or overloaded circuit

fill in: when one fire company is at the scene of a fire for a length of time and another company is sent to their station to handle their calls

fire axe: axe with pick head on one side, regular axe blade on the other side

fire line: the dividing line between where the fire fighters are working and where the spectators are held back by the police

fog nozzles: nozzles that discharge water in the form of small droplets so there is maximum surface contact with the heat, and therefore, maximum cooling

hose jumper: device used to allow traffic to drive across hoses at the scene of a fire.

hose wagon: apparatus used to carry hose and other equipment, but does not have a pump

incendiary: deliberately set

inculpate: to accuse of a crime

indictment: a formal written accusation charging the defendant with a crime

inquest: an official inquiry into a special matter

in service: a unit that is ready to respond to a fire

lay off: to lay the hose off the fire engine from the fire to the water supply

on watch: man assigned to company desk to receive fire alarms

pike pole: pole designed with a metal tip that is pointed so it can be used for pulling objects

post mortem: subsequent to death

post mortem lividity: purplish discoloration of the parts of the body that are nearest the floor. Caused by settling of the blood by gravity into these areas

pyromania: an insane tendency to set fires

rigor mortis: the stiffening of the body muscles after death

roof ladder: short straight ladder with swivel hooks on the tip end. The hooks can be swiveled out and hooked over the peak of the roof

Signal "A": arson; fire of *incendiary* nature

Signal "O": false alarm

spaghetti: large amount of hose on the ground at the scene of a fire

spanner wrench: wrench used for tightening hose connections

static electricity: the charge built by friction on an object

trace evidence: small amount of material which may not be apparent to untrained observer due to its size

under control: when the fire fighters have the fire controlled to the point where there is no danger of it spreading and it can be extinguished with the men and equipment already on the scene.

water thief: an appliance used to permit two $1\frac{1}{2}''$ lines and one $2\frac{1}{2}''$ line to be taken from a single $2\frac{1}{2}''$ supply line

INDEX